Research on Hydraulics and River Dynamics

Research on Hydraulics and River Dynamics

Editor

Vlassios Hrissanthou

MDPI • Basel • Beijing • Wuhan • Barcelona • Belgrade • Manchester • Tokyo • Cluj • Tianjin

Editor
Vlassios Hrissanthou
Democritus University of Thrace
Greece

Editorial Office
MDPI
St. Alban-Anlage 66
4052 Basel, Switzerland

This is a reprint of articles from the Special Issue published online in the open access journal *Water* (ISSN 2073-4441) (available at: https://www.mdpi.com/journal/water/special_issues/Hydraulics_River).

For citation purposes, cite each article independently as indicated on the article page online and as indicated below:

LastName, A.A.; LastName, B.B.; LastName, C.C. Article Title. *Journal Name* **Year**, *Volume Number*, Page Range.

ISBN 978-3-0365-5597-3 (Hbk)
ISBN 978-3-0365-5598-0 (PDF)

Contents

About the Editor

Vlassios Hrissanthou

Vlassios Hrissanthou (Emeritus Professor): Department of Civil Engineering, Democritus University of Thrace, Xanthi, Greece. Study Background: Civil Engineering at the Aristotle University of Thessaloniki, Greece. Postgraduate and doctoral research on Hydrology and Hydraulic Structures at the University of Karlsruhe (KIT), Germany. Postdoctoral research on Hydraulics and Hydraulic Structures at the University of the Armed Forces, Munich, Germany. Teaching Work: (a) Graduate Study Courses: Fluid Mechanics, Hydraulics, Engineering Hydrology, River Engineering, Hydropower Engineering, Water Resources Management, Open Channel Hydraulics, and Hydrology of Groundwater; (b) Postgraduate Study Courses: Advanced Engineering Hydrology, Sediment Transport, Reservoir Design, Time Series Analysis, Selected Chapters of Hydropower Engineering, and Hydraulics of Stratified Flows. Publications: A total of 49 publications in scientific journals, 91 publications in conference proceedings, and 11 publications in book chapters in English, Greek and German. Citations: Overall, 672, excluding self-citations. Reviewer in 50 international scientific journals. Editor of 3 books in English. Author of 1 book in Greek. Guest ditor in 3 Special Issues in English.

water MDPI

Editorial

Research on Hydraulics and River Dynamics

Vlassios Hrissanthou

Department of Civil Engineering, Democritus University of Thrace, 67100 Xanthi, Greece; vhrissan@civil.duth.gr

1. Introduction

River engineering is one of the most important subjects in hydraulic engineering. The main scientific fields that are necessary to understand the basic principles of river engineering include hydrology, hydraulics and geomorphology. Using hydrologic rainfall–runoff models, the river inflows originating from rainfall-induced overland flow can be calculated. In case of intense storms, the extreme situation of flood routing in rivers should be faced. Flood routing can be calculated using both hydrologic and hydraulic models. Hydraulic models are based on the water mass and momentum conservation equations, which are partial differential equations of a hyperbolic type and are solved using numerical methods, e.g., finite difference schemes. The assessment of the hydraulic resistance is a critical point in the hydraulic models.

Soil erosion products from the surrounding basins are transported from the overland flow into the rivers, and constitute the so-called wash load, which is transported in the rivers, in suspension. The riverbed can be eroded by the river flow, or suspended sediment can be deposited on the riverbed. A particularly practical case of bed erosion constitutes the local scour around or downstream of hydraulic structures, e.g., bridge piers or dams, respectively. Generally, the hydraulic structures, e.g., dams, lead to modifications to the hydraulic and geomorphologic regimes in rivers.

Therefore, the geomorphology of riverbeds is strongly affected by the phenomenon of sediment transport. Numerous computational models of bed load and total load have been developed. To consider sediment transport in rivers, the sediment continuity equation should be added to water mass and momentum conservation equations. Sediment transport is mainly influenced by unsteady turbulent flows, which constitute the normal physical condition in rivers. The formation of turbulence and the phenomenon of sediment transport are particularly noticeable in curved channels or river bends. In reservoirs and lakes, the hydraulic and geomorphologic conditions are particularly different from those that dominate in rivers.

In recent years, the flow over a rough bed channel subjected to downward seepage was investigated.

This Special Issue aims to exhibit scientific research on the themes mentioned above.

Citation: Hrissanthou, V. Research on Hydraulics and River Dynamics. *Water* **2022**, *14*, 3018. https://doi.org/10.3390/w14193018

Received: 22 September 2022
Accepted: 23 September 2022
Published: 26 September 2022

2. Overview of This Special Issue

This Special Issue includes nine original contributions focused on river hydraulics. Four of these resulted from cooperation between universities from different countries: (a) Russia and Poland [1], (b) Taiwan and the USA [2], (c) Iran and Italy [3], and (d) India and Italy [4]. The other contributions resulted from research carried out in universities from South Korea [5], Greece [6], China [7,8], and Japan [9].

The nine articles in this Special Issue can be divided into four categories: Category A: "Hydraulic resistance and sediment transport in rivers"; Category B: "Curved channels or river bends"; Category C: "Local scour around spur dikes and pile groups"; Category D: "Rough bed channels subjected to downward seepage". Articles [1,5,6] belong to Category A; Articles [7–9] belong to Category B; Articles [2,3] belong to Category C; Article [4] belongs to Category D.

The article by Gladkov et al. [1] deals with assessments of the hydraulic resistance of the river channel and improvements to the river sediment transport model in a simulation of riverbed transformation on the basis of previous research. To verify the sediment transport model, 296 field measurements of the Central-East European lowland rivers were used. The test calculations show that the modified van Rijn formula provides the best results of all the considered variants.

The article of Sidiropoulos et al. [6] deals with the applicability of the Meyer–Peter and Müller (MPM) bed load transport formula. The performance of the formula is examined based on data collected from a particular location of the Nestos River in Thrace, Greece, in a comparison to a proposed Enhanced MPM (EMPM) formula and two typical machine learning methods, namely, Random Forests (RF) and Gaussian Processes Regression (GPR). The EMPM formula presented a definitely improved performance in comparison to the original formula, which is also competitive with purely data-driven techniques and even superior in the case of smoothed data.

The article by Ahn et al. [5] analyzes a numerical simulation of open-channel turbulent flow over two-dimensional fixed dunes to reveal the effect of roughness on the dune bottom and determine the optimized combination of the turbulence scheme and the roughness height formula. The results of all cases calculated with OpenFOAM are compared with the laboratory experimental data for model validation.

In the study of Zhang et al. [7], the complex flow field in a 60° bend of a river with a groyne is experimentally examined, and the bed shear stress is estimated using the turbulent energy method (TKE method—Tubulent Kinetic Energy method), modified turbulent energy method (TKE–w′ method), Reynolds stress method (Reynolds method), and logarithmic law (log–law) method.

The aim of the study by Andriamboavonju et al. [9] is to experimentally investigate the bedrock incision under sediment bed load transport along curved channels by varying flow speed. Erosion by incision of the bedrock along uniformly curved channels was successfully simulated using the annular flume. The findings of this study reveal that the type of bedform has a more significant impact on bedrock incisions than rotation speed.

According to the article by Zhang et al. [8], laboratory flume experiments were implemented to better understand the influence of transition section configuration on the correlation of flow movement characteristics between the front and back bends of a continuous bend. The three-dimensional instantaneous velocity was measured by Acoustic Doppler Velocimeter (ADV). Through the analysis of circulation structure, circulation intensity and turbulent kinetic energy, the correlation of flow movement characteristics between the front and back bends in a continuous bend under different width/depth ratios of the transition section was studied.

Local scour is a common threat to structures such as bridge piers, abutments and dikes constructed on natural rivers. To reduce the risk of foundation failure, understanding local scour phenomenon around hydraulic structures is important [2]. In the study of Liao et al. [2], a repose angle formula and bed geometry adjustment mechanism are integrated into a 2D mobile-bed model to improve the numerical simulation of local scour holes around structures. A comparison of the calculated and measured bed variation data reveals that a numerical model involving the improvement technique can predict the geometry of local scour holes around spur dikes with reasonable accuracy and reliability.

The study of Daneshfaraz et al. [3] experimentally and numerically investigated the local scour around pile groups positioned upstream and downstream of a harvest pit. Both circular and hydrodynamic shaped piers were considered. Numerical modeling of local scouring around hydrodynamic and circular bridge pile groups was performed using a Computational Fluid Dynamics (CFD) model and Large Eddy Simulation (LES) turbulence model and van Rijn sedimentary model with FLOW-3D software.

The work of Sharma et al. [4] experimentally investigated the turbulent flow characteristics in developing and fully developed flows over a rough bed channel subjected to downward seepage. In concrete terms, the fluid structures in developing and developed wide-open-channel flows over a sandy bed are investigated in the present work. A series of experiments were carried out in a laboratory flume to investigate the fluid features of downward seepage. The streamwise and vertical velocities, as well as the Reynolds shear stress distributions in a developing flow and in a developed flow, were observed and studied in no-seepage and downward seepage conditions.

Funding: This research received no external funding.

Acknowledgments: I acknowledge the contributions of all authors of the nine papers in this Special Issue.

Conflicts of Interest: The author declares no conflict of interest.

References

1. Gladkov, G.; Habel, M.; Babiński, Z.; Belyakov, P. Sediment Transport and Water Flow Resistance in Alluvial River Channels: Modified Model of Transport of Non-Uniform Grain-Size Sediments. *Water* **2021**, *13*, 2038. [CrossRef]
2. Liao, C.-T.; Yeh, K.-C.; Lan, Y.-C.; Jhong, R.-K.; Jia, Y. Improving the 2D Numerical Simulations on Local Scour Hole around Spur Dikes. *Water* **2021**, *13*, 1462. [CrossRef]
3. Daneshfaraz, R.; Ghaderi, A.; Sattariyan, M.; Alinejad, B.; Asl, M.M.; Di Francesco, S. Investigation of Local Scouring around Hydrodynamic and Circular Pile Groups under the Influence of River Material Harvesting Pits. *Water* **2021**, *13*, 2192. [CrossRef]
4. Sharma, A.; Kumar, B.; Oliveto, G. Turbulent Flow Structures in Developing and Fully-Developed Flows under the Impact of Downward Seepage. *Water* **2022**, *14*, 500. [CrossRef]
5. Ahn, J.; Lee, J.; Park, S.W. Optimal Strategy to Tackle a 2D Numerical Analysis of Non-Uniform Flow over Artificial Dune Regions: A Comparison with Bibliography Experimental Results. *Water* **2020**, *12*, 2331. [CrossRef]
6. Sidiropoulos, E.; Vantas, K.; Hrissanthou, V.; Papalaskaris, T. Extending the Applicability of the Meyer-Peter and Müller Bed Load Transport Formula. *Water* **2021**, *13*, 2817. [CrossRef]
7. Zhang, L.; Zhang, F.; Cai, A.; Song, Z.; Tong, S. Comparison of Methods for Bed Shear Stress Estimation in Complex Flow Field of Bend. *Water* **2020**, *12*, 2753. [CrossRef]
8. Zhang, K.; Chen, L.; Li, Y.; Yu, B.; Wang, Y. Experimental Study on the Influence of the Transition Section in the Middle of a Continuous Bend on the Correlation of Flow Movement in the Front and Back Bends. *Water* **2020**, *12*, 3213. [CrossRef]
9. Andriamboavonju, M.R.; Terakado, T.; Izumi, N. Spatiotemporal Evolution of Bed Configurations in Mixed Bedrock-Alluvial in Uniformly Curved Channels. *Water* **2022**, *14*, 397. [CrossRef]

water
MDPI

Article

Turbulent Flow Structures in Developing and Fully-Developed Flows under the Impact of Downward Seepage

Anurag Sharma [1], Bimlesh Kumar [2,*] and Giuseppe Oliveto [3]

1 Department of Civil Engineering, National Institute of Technology Rourkela, Rourkela 769008, India; sharmaan@nitrkl.ac.in
2 Department of Civil Engineering, Indian Institute of Technology Guwahati, Guwahati 781039, India
3 School of Engineering, University of Basilicata, Viale dell'Ateneo Lucano 10, 85100 Potenza, Italy; giuseppe.oliveto@unibas.it
* Correspondence: bimk@iitg.ac.in

Abstract: This work experimentally investigates the turbulent flow characteristics of developing and fully-developed flows over a rough bed channel that is subjected to downward seepage. Instantaneous 3D velocities were collected using an acoustic Doppler velocimeter (ADV) in the developing and fully-developed flow regions, along the channel centerline, to analyze different turbulent statistics. Observations revealed that the streamwise and vertical velocities were higher in developing flows, whereas the Reynolds shear stresses, and turbulence intensities, were smaller. The downward seepage would affect the velocity distributions and flow depth in both the developing and fully-developed regions. Therefore, new equations to represent the distribution of the turbulence intensities were proposed, and a comparison with the current literature is provided. The investigation of the Reynolds stress anisotropy tensors concludes that the degree of anisotropy in fully-developed flows is lower than for developing flows.

Keywords: developing flow; fully-developed flow; Reynolds stress anisotropy tensor; seepage flow; turbulence intensity

Citation: Sharma, A.; Kumar, B.; Oliveto, G. Turbulent Flow Structures in Developing and Fully-Developed Flows under the Impact of Downward Seepage. *Water* **2022**, *14*, 500. https://doi.org/10.3390/w14030500

Academic Editors: Vlassios Hrissanthou and Giuseppe Pezzinga

Received: 22 November 2021
Accepted: 4 February 2022
Published: 8 February 2022

1. Introduction

The flow in open channels is mainly turbulent in nature. In general, open-channel flows are spatially varied in the field due to the frequent variations of the boundary conditions arising from the alterations in flow discharge, as well as other channel parameters [1]. The alterations in boundary conditions cause a flow disruption due to the variance of viscous, gravitational, and inertial forces. Eventually, the flow attempts to reach a fully-developed flow condition in the downstream course. At the channel entrance, a high-velocity gradient is developed close to the channel boundary that is linked with the frictional stresses developed between the fluid particles and the rough surface [2]. The fluid layer close to a rough surface, where viscous properties are apparent, is termed the boundary layer. The boundary layer may be laminar at the inlet section of the channel and gradually thickens up to a definite distance along the channel length, where the flow is termed as developing flow; behind the developing length, the flow becomes a fully-developed flow [3]. For rough boundaries, the viscous properties can be observed in a very thin layer, termed the viscous sublayer, in which the major portion of the velocity variation arises. In fully-developed turbulent flow, with respect to the velocity governing factors, the boundary layer is formed from the inner zone and the outer zone. The inner zone is again composed by the viscous sublayer and the fully turbulent inner region. It is important to investigate the fluid features in developing and developed open-channel flows to evaluate the fluid–sediment interface and sediment aggradation and degradation, which are closely related to the hydrodynamic features, such as the vertical profile of time-averaged velocities, turbulence intensities, Reynolds shear stresses, and turbulent kinetic energy [1]. The turbulent developing open-channel flow is a complex 3D flow governed by a seepage bed, the bed roughness, and other variables.

The two-component flow time-averaged velocity profile in open channels is fundamental in studying bed shear stresses and the occurrence of bed material transport [3–6]. Studies from the literature indicate that the standard log-law is valid for the longitudinal velocity depth profiles in the inner layer of the flow, which is 0.2 times the flow depth above the boundary wall [7,8]. These studies highlight that the log-law diverges from the observed data for the outer layer of flow and is replaced by the wake-law [9]. Kirkgöz and Ardiçlioğlu [2] showed that, in developed flows in wide-open-channels, the flow velocity achieves its maximum value close to the water surface. Mahananda et al. [10] experimentally investigated the turbulence characteristics in developing and fully-developed flows in narrow open-channel flow over a rough bed. They proposed an analytical power law for the time-averaged velocity distribution and matched it with the observed velocity. Besides the laboratory observations, numerous analytical results have also projected the velocity profile in fully-developed flows [11].

There are several studies [4–11] on the turbulence characteristics for fully-developed flows in open channels. On the contrary, the turbulent flow structures in developing flows in open channels are not sufficiently addressed; this is even more evident when seepage conditions also occur. The flow turbulence in the developing zone of an open channel is a three-dimensional flow affected by a downward seepage flow. Studying turbulent features of flow in the developing zone of a channel over a rough bed, which is frequently experienced in fluvial hydraulics, is, therefore, of importance. Marusic et al. [12] specified that a developed flow is the flow in which turbulent parameters and the hydraulic parameters are free from the longitudinal distance towards the flow development. Seepage is a process that influences the turbulent characteristics of the flow and the bed material movement at the bed surface [13,14]. As a practical example, the dynamics of turbulence and seepage govern a significant feature of the transport of solutes through the hyporheic river zone [15], which has important implications for groundwater quality. Field studies [16,17] calculated the quantity of downward seepage in an alluvial river and revealed that the amount of seepage might lie between 10% and 45% of the channel discharge supplied at the canal head. This paper aimed to emphasize the eventual differences between the turbulent flow structures for developing and fully-developed flows in sandy-bed straight channels, considering downward seepage processes. Specifically, the distributions along the flow depth of the streamwise and vertical velocities, turbulence intensities, and Reynolds shear stresses were analyzed. Moreover, the turbulent correlation coefficient, the turbulent kinetic energy (TKE), and the Reynolds stress anisotropy tensor were considered for a deeper exploration.

2. Experimental Methodology

Experiments were conducted in a 17.2 m long, 1.00 m wide, and 0.72 m deep glass-walled rectangular flume in the Water Resources Engineering Laboratory at the Indian Institute of Technology, Guwahati, India. The bed slope of the flume was kept constant for the no-seepage and seepage runs with a value of 0.0005 (Figure 1). A couple of baffle walls were fixed at the inlet of the flume to decrease the degree of flow turbulence when the water entered the channel. The bed was levelled with glass with a smooth finish, and was made with a non-uniform sediment with a median grain size of d_{50} = 0.5 mm and geometric standard deviation of σ_g $(d_{84.1} - d_{15.9})/d_{50}$ = 1.65 to achieve a rough bed surface. It should be noted that $d_{84.1}$, $d_{15.9}$, and d_{50} represented the size of the particles for which the percentage of finer particles were 84.1, 15.9, and 50, respectively. Water was recirculated between the underground trench and the channel by using two centrifugal pumps. The tailgate was adjusted for the required flow depth in the channel. A valve positioned at the upper section of the inlet tank allowed for the preselected flow discharge. Further experimental details are available in Sharma and Kumar [14]. The channel bed was prepared with a non-porous section 2 m from the upstream end of the channel, and the remaining length (15.2 m) was prepared with a porous section by covering it in a fine mesh

(0.1 mm). The fine mesh was sustained by the steel tube arrangement that was 0.22 m high, which was positioned on the bottom of the flume. The experimental setup for downward seepage was prepared by creating a seepage chamber 15.20 m long, 1.0 m wide, and 0.22 m deep. The seepage chamber was located beneath the flume bed to collect the loss of water from the downward seepage. The non-uniform sediment mixture was placed on a fine mesh that was 0.1 mm squared to ensure the free flow conditions from the entry to the seepage zone. The quantity of the seepage discharge was controlled with a pair of valves fixed to the seepage chamber at the outlet end of the flume by forming a pressure difference between the seepage chamber and the channel. The seepage was approximately uniform (slightly non-uniform), and was applied throughout the seepage zone length in order to maintain a constant seepage discharge in the main channel along the streamwise direction. Two electromagnetic flowmeters (EFMs), with an accuracy of $\pm$ 0.5%, were attached to the seepage chamber (pressure chamber) through pipes, which were utilized to measure the desired quantity of the seepage discharge [14]. The uncertainty associated with the experimental data was calculated before performing the experiments. The uncertainty regarding the measurements of the seepage discharge from the electromagnetic flowmeter was analyzed using various repeated tests. The standard uncertainty for the seepage discharge was approximately 3.75%, which suggests a satisfactory accuracy of the EFM used for the discharge measurement. Regarding the coordinate system of the channel, the streamwise velocity component was along the x-axis, the lateral velocity component was along the y-axis, and the vertical velocity component was along the z-axis, which was in a positive direction, upward from the bed surface. The flow depth was measured with a digital point gauge attached to a Vernier scale.

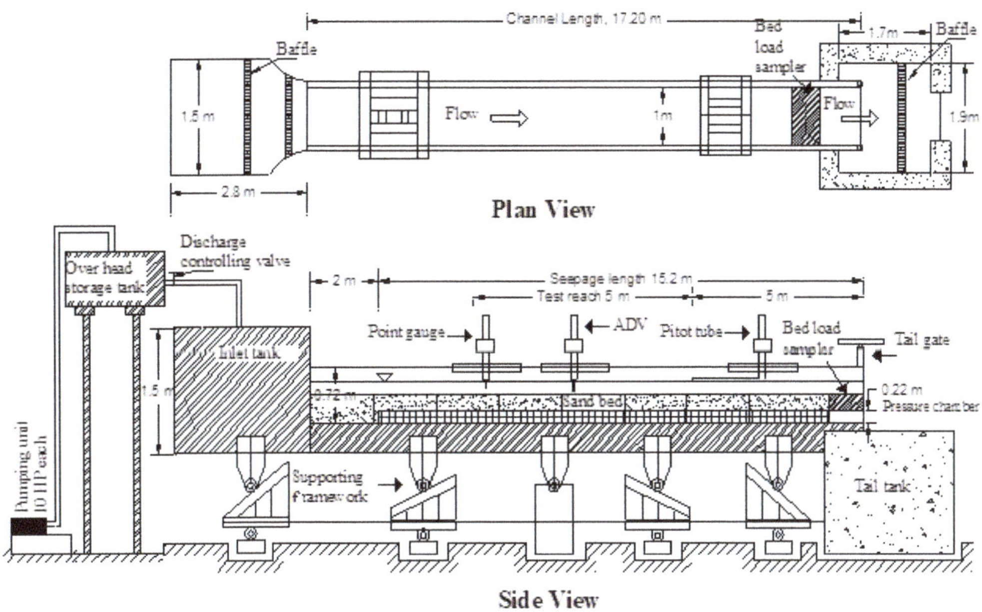

Figure 1. Schematic of experimental setup.

A four-beam down-looking acoustic Doppler velocimeter (VectrinoPlus) was used to measure the three-dimensional instantaneous water velocities. By analyzing the velocity distributions taken along the centerline at every 100 cm, the flow region was observed, fully-developed, when $x \geq 5$ m, consistent with the results of previous studies [3]. Therefore, the velocity data sets were collected at the centerline of the channel cross-section. The data were recorded at various distances from the upstream end of the channel (x = 3 and 7 m) to compare the turbulent flow structures in the developing and fully-developed flows. The nearest measuring point of velocity was taken at a distance of 1–2 mm from the bed surface, and the velocity measurement farthest from the bed was taken 50 mm beneath the water surface. In the vicinity of the bed surface, the velocities were measured with an interval of 1–2 mm, while near the free surface, they were achieved with an interval of 4–5 mm. The instantaneous velocities were measured by ADV at a position 50 mm beneath the probe emitter to reduce the effect of the probe on the experimental data. The data acquisition sampling frequency was 200 Hz. The sampling volume was a 1 to 4 mm long cylinder with a diameter of 6 mm. For the velocity data collection in the flow's outer layer, the sampling volume was maintained at 4 mm long. In the vicinity of channel boundary, the length of the sampling volume was changed to 1 mm so that it did not collide with the sediment particles on the bed. The sampling length of 1 mm was adequate to achieve the real velocity distributions close to the channel bed. The signal-to-noise ratio and the correlation magnitude were higher than 10 and 60, respectively. The instantaneous ADV data were collected for the sampling duration of 300 s. The spikes present in the ADV velocity data were removed by a modified singular spectrum technique, as suggested by Sharma et al. [18].

Two experimental runs were carried out with a flow discharge of Q = 0.0402 m^3/s and seepage discharge of Q_s = 0.00402 m^3/s (Q_s = 10% of the flow discharge). The heights of water for the no-seepage and seepage experiments were 0.116 m and 0.112 m, respectively. The velocity measurements were taken at the developing flow region (x = 3 m from the inlet) and the fully-developed flow region (x = 7 m from the inlet) for the no seepage and downward seepage conditions. These four experimental data sets were recorded as E1 to E4. The experiments E1 and E2 represent the developing and fully-developed flow conditions in the case of the no seepage bed condition. Similarly, the experiments E3 and E4 represent the developing and developed flow in the case of the downward seepage condition. In all experimental runs, the aspect ratio was greater than five [19], ensuring negligible wall effects.

More experimental details are provided in Table 1, with Re, Fr, h, and U representing the flow Reynolds numbers ($Re = Uh/v$, where v is the kinematic viscosity), the flow Froude number $Fr = U/\sqrt{gh}$, the flow depth, and the depth-averaged flow velocity in the developed region of the open-channel flow, respectively.

Table 1. Main hydraulic parameters for the experimental runs of this study.

	h (m)	Q (m^3/s)	Re	U (m/s)	Fr	Q_S (m^3/s)	Experimental Run
No seepage	0.116	0.0402	32,630	0.346	0.3248	0	**E1** (Developing flow) **E2** (Developed flow)
10% seepage	0.112	0.0402	32,850	0.359	0.3424	0.00402	**E3** (Developing flow) **E4** (Developed flow)

3. Results and Discussion

In the present paper, u, v, and w represent the mean streamwise, lateral, and vertical velocity components, respectively, while u', v', and w' are the fluctuating components of the velocities, respectively. Similarly, $\left(\overline{u'u'}\right)^{0.5}$, $(\overline{v\prime v\prime})^{0.5}$ and $(\overline{w\prime w\prime})^{0.5}$ are the turbulence intensities. Moreover, u_* [$=(\tau_0/\rho)^{0.5}$] represents the shear velocity where τ_0 = the boundary shear stress observed from the distribution of the Reynolds shear stresses (τ_{uw}) projected to the bed surface, which is $\tau_0 = (\tau_{uw})_{z=0}$ as given by Nezu and Nakagawa [20]. The shear velocity for the experimental datasets E1, E2, E3, and E4 were 14.41 mm/s, 14.57 mm/s, 14.65 mm/s, and 15.02 mm/s, respectively. In the near-bed zone, the uncertainty errors for the velocities and the Reynolds shear stresses were lower than 5% and 12.5%, respectively, indicating ADV data accuracy. Moreover, based on the flow field classification suggested by Nezu and Nakagawa [20], the present study distinguished the flow field into four layers: (i) the inner layer ($z/h < 0.2$); (ii) the outer layer ($z/h \geq 0.2$); (iii) the intermediate layer

$(0.2 \leq z/h \leq 0.6)$, and (iv) the free surface layer $(0.6 < z/h \leq 1)$. This categorization of the flow field was used to examine the key profiles of the fluid features.

3.1. *Time-Averaged Velocities*

Figure 2 shows the streamwise and vertical velocities against the normalized flow depth (z/h) in the developing and the developed flow regions, where z is the distance from the bed of the measurement point. Velocity distributions, displayed in Figure 2, showed that the position of maximum velocity in the streamwise direction shifted upward, indicating the growth in the boundary layer thickness along the developing region [20,21]. The data of the experimental runs E1 and E2 did not overlap, but the velocity distributions were distinct for the developing and developed flows. In the case of the no-seepage runs (i.e., runs E1 and E2), the mean velocity of the streamwise and vertical directions for the developing flow fulfilled its maximum value in the vicinity of the water surface and slowly decreased towards the channel boundary because of the flow resistance due to the bed roughness. In the condition of the developing flows, the velocity values in the streamwise and vertical directions were amplified in comparison to the developed flow. The flow depth was slightly lower in the developing zone, and gradually increased along the channel length until the fully-developed flow condition was achieved. As a consequence, the streamwise and vertical flow velocities were higher in the developing flow. Measurements revealed that the streamwise velocities close to the channel boundary increased by approximately 7–9% in the developing flow, and the vertical velocities increased by approximately 10–12%. The streamwise and vertical velocity profiles for developing turbulent boundary layers are in line with the available literature [10,21]. In the seepage experiments, velocity measurements through ADV were taken immediately after the application of seepage to understand the effects of downward seepage on the flow characteristics of the channel. Flow velocities in both the developing and developed flows were higher than in the no-seepage conditions. Similar to the no-seepage conditions, in the developing flow, the streamwise velocities close to the channel bed increased by approximately 6–8%, and the vertical velocities increased by approximately 9–12%, in comparison to the developed flow. The velocity defect in the outer zone was more in the developing flow condition, as compared to the developed flows. It was observed that seepage flow influenced the vertical velocities (w) more than those of the streamwise velocities (u). The observed data nearly overlapped for the streamwise velocity, whereas the experimental data were far apart in the case of the vertical velocity, indicating a higher level of turbulence with seepage. The existing literature in the application of downward seepage shows that the distribution of the velocity tends to move vertically downward and, therefore, a greater velocity is observed close to the channel boundary [22].

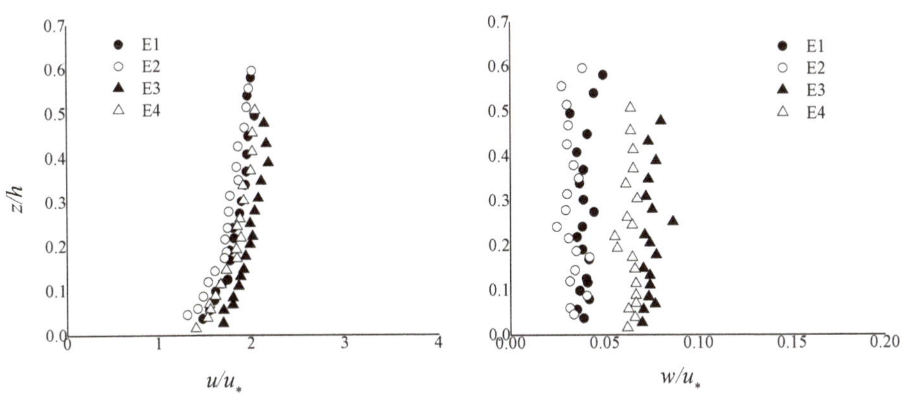

Figure 2. Vertical distribution of streamwise (u) and vertical (w) velocities for runs **E1** (developing flow and no-seepage), **E2** (fully-developed flow and no-seepage), **E3** (developing flow and downward seepage), and **E4** (fully-developed flow and downward seepage).

3.2. *Reynolds Shear Stresses (RSSs)*

The momentum exchange between the fluid structure and the bed material in the channel is called Reynolds shear stress. Hence, the RSS is a key factor that leads to sediment transport and erosion in the channel. The distributions of RSSs along the flow depth for the developing and developed flows over the no-seepage and downward seepage beds are presented in Figure 3. It was observed that RSS increased towards the bed, which is related to the momentum delivered from the fluid structures to the channel boundary, thus promoting sediment transport and overcoming the resistance of the grain. This observation is in agreement with previous studies [20,23]. The RSS achieved a peak magnitude in the inner layer of flow, and it reduced close to the channel boundary due to the roughness sub-layer. Figure 3 shows similar data trends for the RSS distributions in the developing and developed flows, but with different magnitudes. The greater magnitudes of RSS were found for the developed flow in comparison with the developing flow. This would imply the presence of a higher momentum exchange in developed flows. In addition, the upward decreasing trend of RSS would indicate that the farther the measurement point is from the channel boundary, the lower the turbulence generated by the flow in the developing and developed regions is. Hence, the distributions of RSS were evaluated to describe the momentum diffusion phenomena. A significant finding was that the peak value of RSS in the developed flow was higher than that in the developing flow by approximately 22.4% and 6.7% for the no-seepage and downward seepage conditions, respectively. In all the distributions, amplified values of RSSs were observed for runs in which downward seepage processes occurred. This would imply a greater momentum exchange near the bed, thus increasing the sediment mobility in the channel. In addition, a substantial increase in the RSS values were ascertained in the developed flows, resulting in higher velocity fluctuations in comparison with the developing flow. RSS indicated the turbulent intensities in the flow, as it was dependent on the velocity fluctuations.

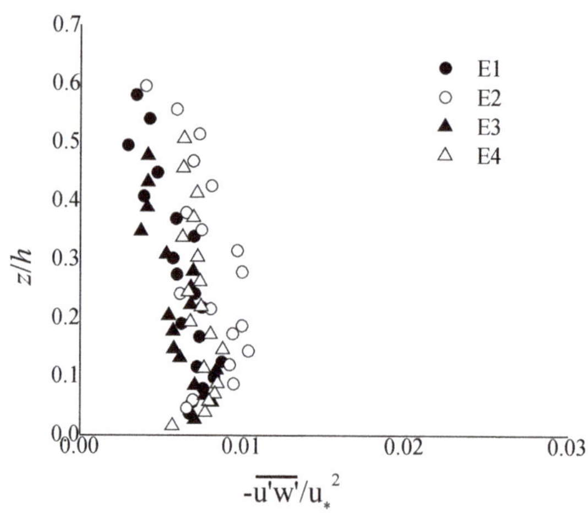

Figure 3. Depth profile of the Reynolds shear stress for runs **E1** (developing flow and no-seepage), **E2** (fully-developed flow and no-seepage), **E3** (developing flow and downward seepage), and **E4** (fully-developed flow and downward seepage).

3.3. *Turbulence Intensities and Turbulence Kinetic Energy (TKE)*

The turbulence intensity is an outcome of the variation in the instantaneous velocity at a measurement point determined by the velocity measurement and the type of bed roughness. Additionally, the root mean square (rms) of the fluctuating component of the velocity indicates the turbulence intensities. Hence, the streamwise and vertical turbulence intensities represented by σ_u, and σ_w, respectively, are given by

$$\sigma_u = \sqrt{\frac{\sum_{i=1}^{n}(U_i - u)^2}{n}} \quad (1)$$

$$\sigma_w = \sqrt{\frac{\sum_{i=1}^{n}(W_i - w)^2}{n}} \quad (2)$$

where n is the number of samples, U_i and W_i represent the instantaneous velocities in the streamwise and vertical directions, respectively, and u and w represent the mean streamwise and vertical velocity components, respectively. The distributions along the flow depth of the turbulence intensities for the developing and developed flows are presented in Figure 4. The peak value for turbulence intensity was attained in the inner flow layer in both the developing and developed flows, where the maximum value of RSS was also observed with a tendency towards a steady condition. Close to the free surface, the turbulence intensities tended towards stationary values. Fluctuations in values were more pronounced near the channel boundary than in the region close to the water surface, due to the bed roughness. In addition, the degree of the turbulence intensity was higher in the developed flows than in the developing flows. A significant finding was that the near-bed streamwise turbulence intensities (σ_u) in the developed flow were higher by 10–30% than those in the developing flows over the seepage bed, whereas the increasing magnitude for the near-bed vertical turbulence intensities (σ_w) was 5–10%. The downward seepage in the channel escalated the flow velocity fluctuations in the near-bed region, with an increase in the flow turbulence near the bed. It was also observed that the distributions of the turbulence intensities over the seepage bed were higher in near-bed region, though slightly, compared to those in the case of the no-seepage bed condition. This result is in line with the existing literature [24,25].

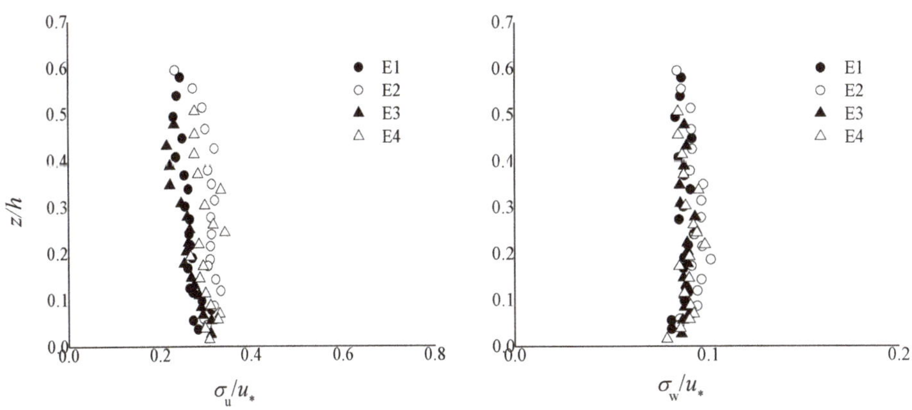

Figure 4. Vertical distribution of streamwise and vertical turbulence intensity for runs **E1** (developing flow and no-seepage), **E2** (fully-developed flow and no-seepage), **E3** (developing flow and downward seepage), and **E4** (fully-developed flow and downward seepage).

The normalized form of the streamwise (σ_u) and vertical (σ_w) turbulence intensities are represented by $\widehat{\sigma}_u = \sigma_u / u_*$ and $\widehat{\sigma}_w = \sigma_w / u_*$, where u_* is the shear velocity. An effort was made to revise the literature equations proposed for wide-open-channel flows. Numerous models were recommended to predict the distributions of the turbulence intensities and the flow depths in wide channels [1,6,17,20,26,27]. The existing literature models have not considered the developing flow over the seepage bed, and the trends for the streamwise and vertical turbulence intensities in the developing and developed flows for the seepage experiments presented in Figures 5 and 6 do not match the universal models provided by previous studies. A regression study was performed to develop new empirical equations for flows in the developing and developed zones for no-seepage and downward seepage conditions. The functional forms of the semi-empirical expressions suggested in the literature [20,26] is considered in the present work, as shown below:

$$\left.\begin{array}{l}\frac{\sigma_u}{u_*}=\alpha_1\exp\left(-\beta_1\frac{z}{h}\right)\\ \frac{\sigma_w}{u_*}=\alpha_2\exp\left(-\beta_2\frac{z}{h}\right)\end{array}\right\}\text{developing flow}\qquad \left.\begin{array}{l}\frac{\sigma_u}{u_*}=\alpha_3\exp\left(-\beta_3\frac{z}{h}\right)\\ \frac{\sigma_w}{u_*}=\alpha_4\exp\left(-\beta_4\frac{z}{h}\right)\end{array}\right\}\text{developed flow} \tag{3}$$

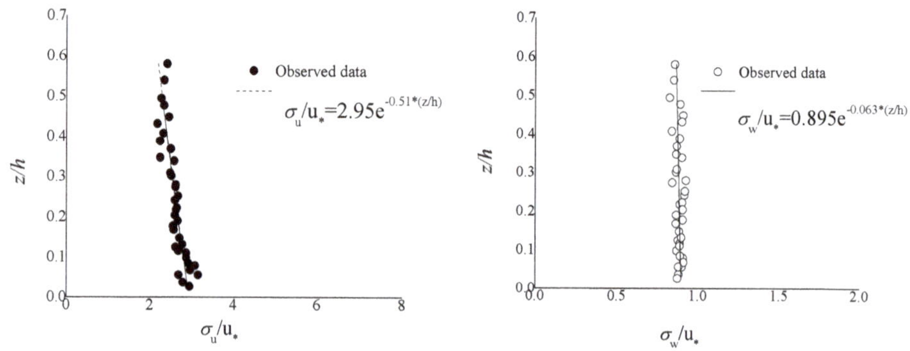

Figure 5. Vertical distributions of normalized streamwise and vertical turbulence intensities in developing flow.

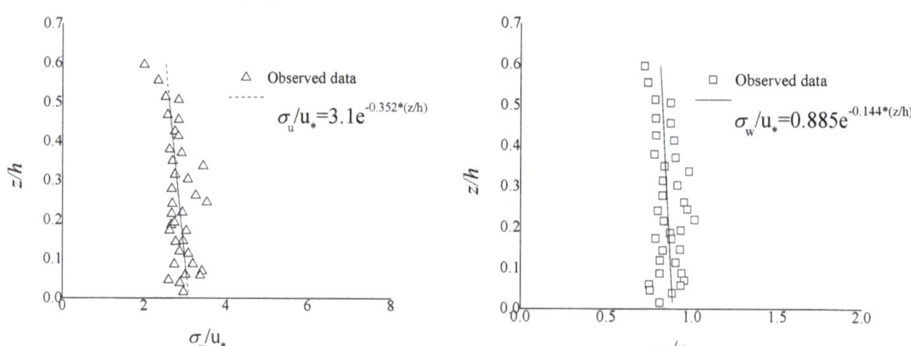

Figure 6. Vertical distributions of normalized streamwise and vertical turbulence intensities in developed flow.

In the above equations, the coefficients α_1, α_2, α_3, and α_4, and the exponential coefficients β_1, β_2, β_3, and β_4 were found from the data collected in this study. In particular, a regression study was performed. The current work revealed a modification in the magnitude of the coefficient and the exponential values from existing studies. The coefficients suggested here, as well as those from the existing literature are given in Table 2. The proposed empirical expressions are also displayed in Figures 5 and 6. The previous studies [20,28] remarked that the coefficient (α_3) of the streamwise turbulent intensities equation was greater than that (α_4) for the vertical turbulence intensities, i.e., $\alpha_3 > \alpha_4$; the same result was also found in the current work, as shown in Table 2. The velocity fluctuations in the streamwise direction would increase the streamwise turbulence intensities, whereas the water surface was prone to decrease the vertical turbulence intensity in the vertical direction. The evaluation of the coefficients in the current work, and those from previous studies for smooth and rough beds, highlighted that the coefficient for the developed flows over the seepage bed was higher, indicating that the turbulence intensity profiles, in the case of seepage, were altered in comparison with those for wide-open-channel flows.

Table 2. Values of the coefficients observed from turbulent intensity equations.

Literature	α_1	α_2	α_3	α_4	β_1	β_2	β_3	β_4	Bed Characteristics
Present study	2.95	0.895	3.1	0.88	0.51	0.063	0.352	0.144	Hydraulically rough and seepage bed
Cardoso et al. [6]	-	-	2.28	-	-	-	1.08	-	Hydraulically smooth
Kironoto and Graf [17]	-	-	2.04	1.14	-	-	0.97	0.76	Hydraulically rough
Nezu and Azuma [26]	-	-	2.30	1.27	-	-	1	1	Hydraulically smooth and rough
Nezu and Nakagawa [20]	-	-	2.30	1.27	-	-	1	1	Hydraulically smooth and rough
Nezu and Rodi [27]	-	-	2.26	1.23	-	-	1	1	Hydraulically smooth and rough
Mahananda et al. [1]	-	-	2.52	1.35	-	-	1	1	Hydraulically rough and narrow open-channel flow

In addition, the distributions, along with the flow depth of the turbulent kinetic energy (TKE), $K_{2D} = 0.75\left(\overline{u'^2} + \overline{w'^2}\right)$ for the developing and developed flows, and in the cases of the no-seepage and downward seepage conditions, are displayed in Figure 7. It can be noted that for the developing and developed flows, the distributions did not overlap in both the no-seepage and downward seepage conditions. For a given flow discharge, the seepage run exhibited a greater mean TKE in the vicinity of the bed surface than the no-seepage run. A greater magnitude of TKE in the vicinity of the bed surface was observed for the developed flows due to the greater streamwise and vertical turbulence intensities [10]. As in the case of the fully-developed flows, the distribution of TKE in the developing flows showed the highest magnitude close to the channel boundary, and reduced quickly in the outer flow region. Finally, the vertical profiles of TKE for a given flow discharge were amplified for the developed flow, in comparison with the developing flows.

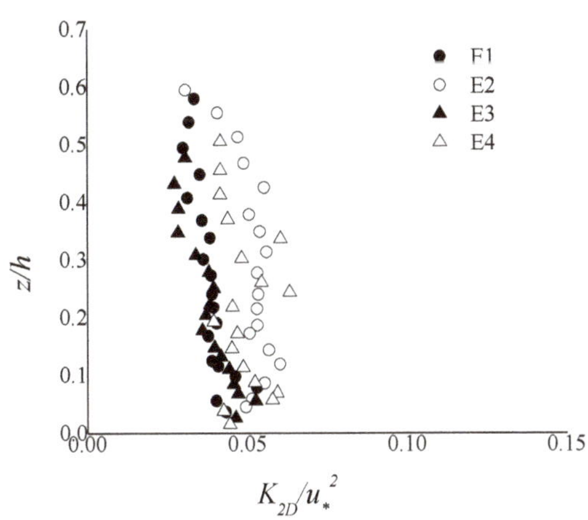

Figure 7. Vertical distribution of TKE for runs **E1** (developing flow and no-seepage), **E2** (fully-developed flow and no-seepage), **E3** (developing flow and downward seepage), and **E4** (fully-developed flow and downward seepage).

3.4. Fluctuating Velocities

Lu and Willmarth [29] developed the quadrant analysis to examine the presence of a coherent structure and to compute the input of coherent eddies into the production of Reynolds shear stresses. In this study, the velocity fluctuations in the four quadrants were explored in the flow's inner layer for the developing and developed flows to examine the comparative role of bursting events in producing turbulent shear stresses. In the quadrant analysis, streamwise u′ and vertical w′ fluctuating velocities were distributed into four quadrants on a u′-w′ plane. Each quadrant represented a type of bursting event, which occurred in the flow zone. The first quadrant represented outward interactions (u′ > 0, w′ > 0), which specified the outward transport of the high-speed fluid parcels. The second quadrant represented the ejections events (u′ < 0, w′ > 0), which implied the drive away of the low-speed fluid parcel from the channel boundary. Inward interactions (u′ < 0, w′ < 0), which defined low-speed fluid parcel motion towards the channel boundary, were signified by the third quadrant. The fourth quadrant (u′ > 0, w′ < 0) characterized sweep events defined by high-speed fluid parcel movement towards the channel bed. For the quadrant analysis, the position $z = 0.05h$ in the inner region was selected as a measurement point for the developing and developed flows for all experimental runs. For the developing flow, Figure 8 shows the scatter variations of the fluctuating velocities u′ and w′ for a hole size $H = 0$ in the case of no seepage (the diagram on the left) and downward seepage (the diagram on the right), respectively. The hole size $H = 0$ indicated the high frequency events, including the small values associated with the use of all the pairs of u′ and w′. By comparing the scatter plots for hole size $H = 0$, it was found that the level of the bursting events increased in the developed flow zone for the no-seepage and seepage conditions. A further important observation was that more events took place in the second and fourth quadrants in the vicinity of bed surface ($z = 0.05h$), showing that the sweep and ejection events were prone to govern the flow. These results are in line with the observations made by Bomminayuni and Stoesser [30]. From the comparison of the two data plots, it can be observed that the levels of the bursting events were stronger for the developing flows in the presence of downward seepage. With seepage, in the case of the developing flow, the average contributions from ejections and sweeps are increased by 28.6% and 24.5%, respectively, in comparison to the no seepage value.

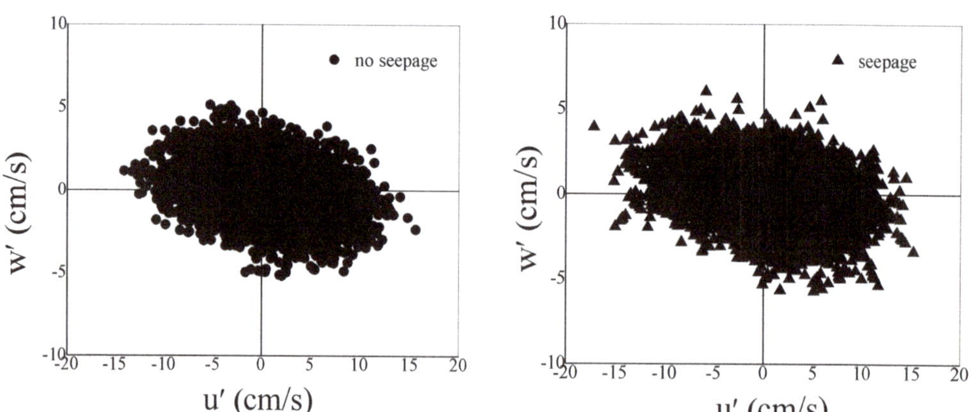

Figure 8. u′-w′ scatter plots with hole size $H = 0$ at selected point $0.05h$ in the developing flow region.

Figure 9 shows the scatter variations in the streamwise u′ and vertical w′ fluctuating velocities for hole size $H = 0$ in the inner regions of the developed flow, in the cases of the no-seepage and downward seepage experiments. The hole size $H = 0$ indicated that all u′ and w′ velocities were considered at the measurement point ($z = 0.05h$). Figure 9 shows that the sweep and ejection events mainly characterized the bursting events in the inner regions for the no-seepage and downward seepage conditions, while the contribution of the outward and inward interactions was somewhat weak, as compared to the sweeps and ejections through the entire flow depth. The comparison of the scatter variation of streamwise u′ and vertical w′ showed that the degree of the bursting events was higher in second and fourth quadrants (ejection and sweep). In the no-seepage and downward seepage conditions, the plot area of the velocity fluctuations in the inner region was nearly eclipsed, which showed that the turbulence was highly anisotropic. The degree of turbulence fluctuations without normalization increased with seepage, as can be seen in Figure 9. The input of bursting events to the production of Reynolds stress, as a measurement point, can be expressed as [20]

$$\langle u\prime w\prime \rangle_{i,H} = \lim_{T\to\infty} \frac{1}{T} \int_0^T u\prime(t) w\prime(t) I_{i,H}[u\prime(t) w\prime(t)] \mathrm{d}t \quad (4)$$

where t, T, and $I_{i,H}$ signify the time, sampling duration, and the indicator function, respectively. The indicator function is expressed as

$$I_{i,H}[u'(t)w'(t)] = \begin{cases} 1, \text{ if } (u', w') \text{is in quadrant } i \text{ and if } |u'w'| \geq H\left(\overline{u'^2}\right)^{0.5}\left(\overline{w'^2}\right)^{0.5} \\ 0, \text{ otherwise} \end{cases} \quad (5)$$

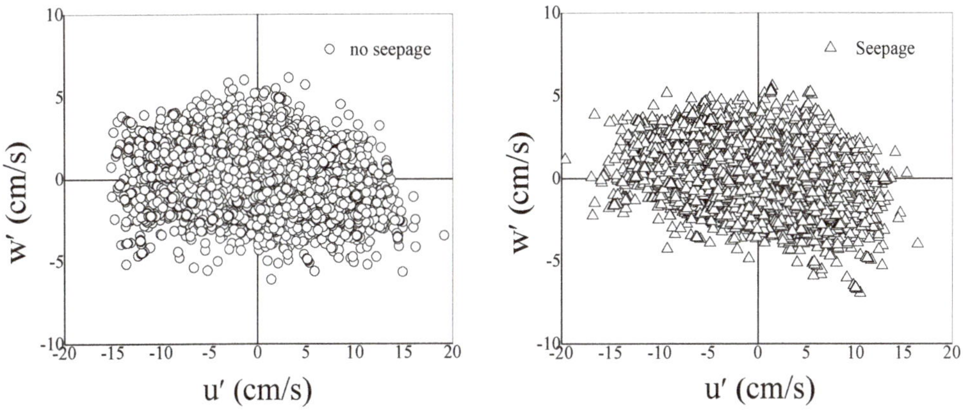

Figure 9. u′-w′ scatter plots with hole size $H = 0$ at a depth $z = 0.05h$ in fully-developed flow region.

A greater number of events occurred in the second and fourth quadrants, describing the ejections and sweeps, which governed the flow. With seepage, in the case of developed flow, the average contributions from ejections and sweeps were increased by 13.6% and 3.5%, respectively, in comparison to the no-seepage condition. From Figures 8 and 9, and for a given flow discharge, the developed flow over the seepage bed had the maximum fluctuations, while the developing flow with the no-seepage bed was characterized by the lowest fluctuations.

3.5. Turbulent Correlation Coefficient

Figure 10 presents the depth profiles of the correlation coefficient $r_{uw} = \left[-\overline{u'w'}/\sigma_u\sigma_w\right]$ (at different water heights) for the developing and fully-developed flows with no-seepage and downward seepage conditions. It was observed that the correlation coefficient for the fully-developed flow was typically higher than that of the developing flow throughout the flow depth. The change in the correlation coefficient for the fully-developed flow was due to the relative modification of the Reynolds shear stress over the turbulence intensity. Figure 10 also reveals that the correlation coefficient reduces, although very gradually, in the vicinity of the bed surface for both the developing and fully-developed flows. Interestingly, this observation matches the findings of Kironoto and Graf [31] for hydraulically rough wide-channels. The effect of seepage on the correlation coefficient was noticeable, as the correlation coefficient near the bed for the seepage runs was generally lower, by 5–10%, than in the case of the no-seepage condition. The observed correlation coefficient changes over the seepage bed were due to the comparative alteration of the Reynolds shear stresses over the turbulence intensities.

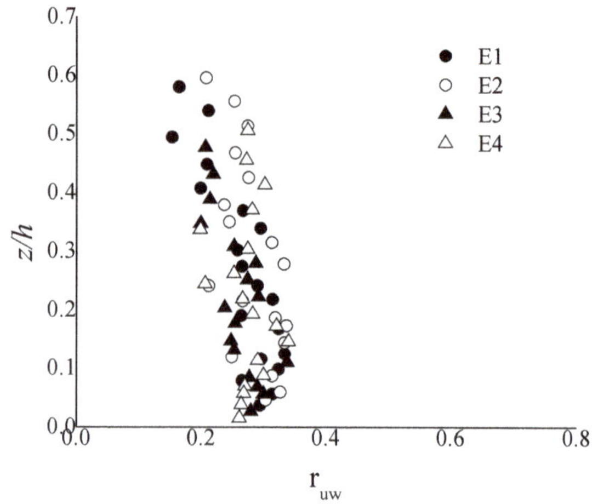

Figure 10. Vertical distribution of the correlation coefficient r_{uw} (i.e., the ratio of the Reynolds shear stress to the product of turbulence intensities in streamwise and vertical directions) for runs **E1** (developing flow and no-seepage), **E2** (fully-developed flow and no-seepage), **E3** (developing flow and downward seepage), and **E4** (fully-developed flow and downward seepage).

3.6. Reynolds Stress Anisotropy Tensor

Reynolds stress anisotropy proposed the approximation for the level of deviation from the isotropy in order to calculate the anisotropy in a turbulent flow. Therefore, the fluctuating components of velocities were used to evaluate the anisotropy tensor of the Reynolds shear stress, which further illustrated that the flow was anisotropic in nature. The Reynolds stress anisotropy tensor b_{ij} was expressed as the ratio of the Reynolds stress parameter to the turbulence kinetic energy, minus its isotropic equivalent magnitude. Reynolds stress anisotropy is an estimate of the deviation from isotropic turbulence (b_{ij} = 0). Thus, b_{ij} is calculated as $b_{ij} = \frac{\overline{u_i'u_j'}}{2K} - \frac{\delta_{ij}}{3}$ where $K = \frac{1}{2}\left(\overline{u'u'} + \overline{v'v'} + \overline{w'w'}\right)$ represents the turbulent kinetic energy, and δ_{ij} denotes the Kronecker delta function, defined by δ_{ij} $(i \neq j) = 0$ and δ_{ij} $(i = j) = 1$. From a physical point of view, the anisotropy tensor components b_{11}, b_{22}, and b_{33} can be regarded as the relative contributions of the turbulence intensities in streamwise, spanwise, and vertical directions to the average turbulent kinetic energy. On the other hand, the component b_{13} signifies the ratio of the Reynolds shear stress to the average turbulent kinetic energy value. The anisotropy parameters b_{11}, b_{12}, b_{22}, and b_{33} are plotted in Figure 11. The parameter b_{ij} is a symmetric and traceless tensor, given as $1/3 \leq b_{ij} \leq 2/3$, and b_{ij} disappears in isotropic turbulence. The symbol of each diagonal variable in b_{ij} signifies the respective input of the turbulence intensity parameters, compared with the turbulent kinetic energy. Figure 11 shows the changes in the stress anisotropy tensors b_{ij} with a non-dimensional flow depth z/h for the developing and developed flows, where z represents the vertical distance taken from the bed surface, as mentioned above. The profile of the anisotropy tensor component b_{11} for the developing flow implied the higher anisotropic stream at the near-bed region in the flow direction in comparison to that of b_{11} for developed flow, whereas b_{13}, b_{22}, and b_{33} in the vicinity of the bed level for developing flow provided the lower anisotropic stream in comparison to those of b_{13}, b_{22}, and b_{33} of the developed flow. Papanicolaou et al. [32] also observed that the features of bed morphology influenced the fluid Reynolds stresses; therefore, the current results are in good agreement with the existing literature. Further, the anisotropy parameters b_{13}, b_{22}, and b_{33} can be considered the comparative influence of the turbulence intensities in streamwise, spanwise, and vertical directions on the average turbulent kinetic energy.

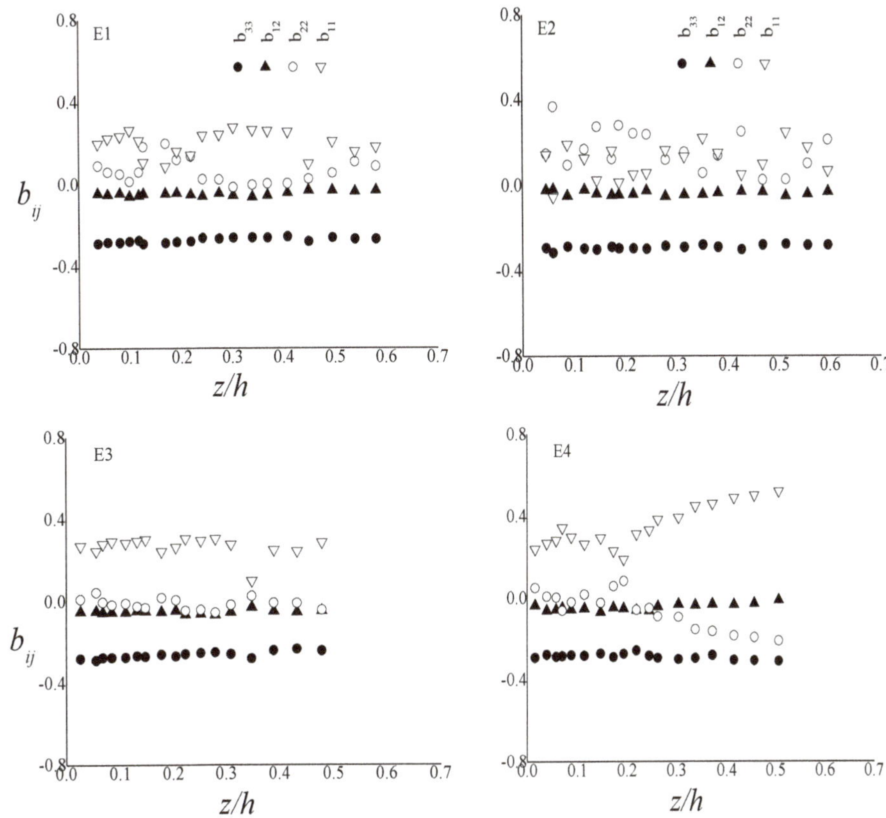

Figure 11. Reynolds stress anisotropy tensor (b_{33}, b_{12}, b_{22} and b_{11}) for runs **E1** (developing flow and no-seepage), **E2** (fully-developed flow and no-seepage), **E3** (developing flow and downward seepage), and **E4** (fully-developed flow and downward seepage).

4. Conclusions

The fluid structures in the developing and developed wide-open-channel flows over a sandy bed were investigated in the present work. A series of experiments were carried out in a laboratory flume to investigate the fluid features of downward seepage. The streamwise and vertical velocity variations in the developing flow were found to be higher than that in the developed flow. Similar profiles were observed in the no-seepage and downward seepage conditions, but streamwise and vertical velocities increased under the influence of seepage. The Reynolds shear stress (RSS) distributions were lightly scattered and increased in the developed flow, implying a larger momentum transfer from the flow to the channel bed. Similar distributions were found in the no-seepage and seepage runs, where RSS also increased under the effect of seepage. The profiles of the RSS follow the damping trends in the vicinity of the channel boundary because of the decreased velocity fluctuations near the channel bed. In the developed flow region, the streamwise and vertical turbulence intensities were found to be amplified compared to the developing flow. New exponential expressions were developed for turbulence intensities in the developing and developed flows over the seepage bed. The streamwise anisotropy tensor in the developing flow depicts the higher values of an anisotropic stream in the vicinity of the bed surface, compared to the developed flow. The quadrant analysis results signified that the level of instantaneous velocity fluctuations, without normalization, increased in the developed flow for a given flow discharge. Further, the instantaneous velocity fluctuations in the developing flow decreased in the no seepage and downward seepage conditions. For all experimental runs, the turbulent correlation coefficient increased in the developed flow.

Author Contributions: For conceptualization, A.S.; methodology, A.S. and B.K.; formal analysis, A.S. and B.K.; investigation, A.S. and B.K.; writing—original draft preparation, A.S.; writing—review and editing, A.S., B.K. and G.O.; visualization, G.O. All authors have read and agreed to the published version of the manuscript.

Funding: This research received no external funding.

Institutional Review Board Statement: Not applicable.

Informed Consent Statement: Not applicable.

Data Availability Statement: The data presented in this study are available on request from the corresponding author.

Conflicts of Interest: The authors declare no conflict of interest.

References

1. Mahananda, M.; Hanmaiahgari, P.R.; Balachandar, R. Effect of aspect ratio on developing and developed narrow open channel flow with rough bed. *Can. J. Civ. Eng.* **2018**, *45*, 780–794. [CrossRef]
2. Kirkgöz, M.S.; Ardiçlioğlu, M. Velocity profiles of developing and developed open channel flow. *J. Hydraul. Eng.* **1997**, *123*, 1099–1105. [CrossRef]
3. Sharma, A.; Kumar, B. Boundary layer development over non-uniform sand rough bed channel. *ISH J. Hydraul. Eng.* **2019**, *25*, 162–169. [CrossRef]
4. Maclean, A.G. Open channel velocity profiles over a zone of rapid infiltration. *J. Hydraul. Res.* **1991**, *29*, 15–27. [CrossRef]
5. Shahiri Tabarestani, E.; Afzalimehr, H.; Pham, Q.B. Validation of double averaged velocity method in a variable width river. *Earth Sci. Inform.* **2021**, *14*, 2265–2278. [CrossRef]
6. Tominaga, A.; Nezu, L. Velocity profiles in steep open-channel flows. *J. Hydraul. Eng.* **1992**, *118*, 73–90. [CrossRef]
7. Cardoso, A.H.; Graf, W.H.; Gust, G. Uniform flow in a smooth open channel. *J. Hydraul. Res.* **1989**, *27*, 603–616. [CrossRef]
8. Xinyu, L.; Changzhi, C.; Zengnan, D. Turbulent flows in smooth-wall open channels with different slope. *J. Hydraul. Res.* **1995**, *33*, 333–347. [CrossRef]
9. Coles, D. The law of the wake in the turbulent boundary layer. *J. Fluid Mech.* **1956**, *1*, 191–226. [CrossRef]
10. Mahananda, M.; Hanmaiahgari, P.R.; Balachandar, R. On the turbulence characteristics in developed and developing rough narrow open-channel flow. *J. Hydro Environ. Res.* **2021**, *40*, 17–27. [CrossRef]
11. Han, Y.; Yang, S.-Q.; Sivakumar, M.; Qiu, L.-C. Investigation of velocity distribution in open channel flows based on conditional average of turbulent structures. *Math. Probl. Eng.* **2017**, *2017*, 1458591. [CrossRef]
12. Marusic, I.; McKeon, B.J.; Monkewitz, P.A.; Nagib, H.M.; Smits, A.J.; Sreenivasan, K.R. Wall-bounded turbulent flows at high Reynolds numbers: Recent advances and key issues. *Phys. Fluids* **2010**, *22*, 065103. [CrossRef]
13. Rao, A.R.; Sreenivasulu, G.; Kumar, B. Geometry of sand-bed channels with seepage. *Geomorphology* **2011**, *128*, 171–177. [CrossRef]
14. Sharma, A.; Kumar, B. Structure of turbulence over non uniform sand bed channel with downward seepage. *Eur. J. Mech. B Fluid.* **2017**, *65*, 530–551. [CrossRef]
15. Liu, Y.; Wallace, C.D.; Zhou, Y.; Ershadnia, R.; Behzadi, F.; Dwivedi, D.; Xue, L.; Soltanian, M.R. Influence of streambed heterogeneity on hyporheic flow and sorptive solute transport. *Water* **2020**, *12*, 1547. [CrossRef]
16. ANCID. *Australian Irrigation Water Provider Benchmarking Data Report for 2004–2005*; Australian National Committee on Irrigation and Drainage: Canberra, Australia, 2006.
17. Tanji, K.K.; Kielen, N.C. *Agricultural Drainage Water Management in Arid and Semi-Arid Areas*; FAO Irrigation and Drainage Paper 61, Publishing Management Service, Information Division; FAO: Rome, Italy, 2002.
18. Sharma, A.; Maddirala, A.K.; Kumar, B. Modified singular spectrum analysis for despiking acoustic Doppler velocimeter (ADV) data. *Meas. J. Int. Meas. Confed.* **2018**, *117*, 339–346. [CrossRef]
19. Martin, C.A.; Gates, T.K. Uncertainty of canal seepage losses estimated using flowing water balance with acoustic Doppler devices. *J. Hydrol.* **2014**, *517*, 746–761. [CrossRef]
20. Nezu, I.; Nakagawa, H. *Turbulence in Open-Channel Flows*; IAHR Monograph, A.A. Balkema: Rotterdam, The Netherlands, 1993.

21. Lane, S.N.; Biron, P.M.; Bradbrook, K.F.; Butler, J.B.; Chandler, J.H.; Crowell, M.D.; Roy, A.G. Three-dimensional measurement of river channel flow processes using acoustic Doppler velocimetry. *Earth Surf. Process. Landf.* **1998**, *23*, 1247–1267. [CrossRef]
22. Cao, D.; Chiew, Y.M. Suction effects on sediment transport in closed conduit flows. *J. Hydraul. Eng.* **2014**, *140*, 04014008. [CrossRef]
23. Bonakdari, H.; Lipeme-Kouyi, G.; Asawa, G.L. Developing turbulent flows in rectangular channels: A parametric study. *J. Appl. Res. Water Wastewater* **2014**, *1*, 51–56.
24. Faruque, M.A.A.; Balachandar, R. Seepage effects on turbulence characteristics in an open channel flow. *Can. J. Civ. Eng.* **2011**, *38*, 785–799.
25. Lu, Y.; Chiew, Y.M.; Cheng, N.S. Review of seepage effects on turbulent open-channel flow and sediment entrainment. *J. Hydraul. Res.* **2008**, *46*, 476–488. [CrossRef]
26. Nezu, I.; Azuma, R. Turbulence characteristics and interaction between particles and fluid in particle-laden open channel flows. *J. Hydraul. Eng.* **2004**, *130*, 988–1001. [CrossRef]
27. Nezu, I.; Rodi, W. Open-channel flow measurements with a Laser Doppler Anemometer. *J. Hydraul. Eng.* **1986**, *112*, 335–355. [CrossRef]
28. Papanicolaou, A.N.; Hilldale, R. Turbulence characteristics in gradual channel transition. *J. Eng. Mech.* **2002**, *128*, 948–960. [CrossRef]
29. Lu, S.S.; Willmarth, W.W. Measurements of the structure of the Reynolds stress in a turbulent boundary layer. *J. Fluid Mech.* **1973**, *60*, 481–511. [CrossRef]
30. Bomminayuni, S.; Stoesser, T. Turbulence statistics in an open-channel flow over a rough bed. *J. Hydraul. Eng.* **2011**, *137*, 1347–1358. [CrossRef]
31. Kironoto, B.A.; Graf, W.H. Turbulence characteristics in rough uniform open-channel flow. *Proc. Inst. Civ. Eng. Water Marit. Eng.* **1994**, *106*, 333–344. [CrossRef]
32. Papanicolaou, A.N.; Diplas, P.; Evaggelopoulos, N.; Fotopoulos, S. Stochastic incipient motion criterion for spheres under various bed packing conditions. *J. Hydraul. Eng.* **2002**, *128*, 369–380. [CrossRef]

water

MDPI

Article

Spatiotemporal Evolution of Bed Configurations in Mixed Bedrock-Alluvial in Uniformly Curved Channels

Mamy Rija Andriamboavonjy *, Tomoya Terakado and Norihiro Izumi

Division of Field Engineering for the Environment, Graduate School of Engineering, Hokkaido University, Sapporo 060-8628, Japan; tera135791@eis.hokudai.ac.jp (T.T.); nizumi@eis.hokudai.ac.jp (N.I.)
* Correspondence: andrymamyrija@gmail.com or andrymamyrija2@eis.hokudai.ac.jp; Tel.: +81-70-2427-8027

Abstract: River courses are rarely straight. Rather, they tend to be meandering. Incision meandering is one of the most common types of meandering discussed in river engineering. The availability of abrasive tools, coverage thickness, appropriate flow velocity, channel geometry, and flow level play a role in the natural phenomenon of bedrock incisions. Any minor change in those parameters, whether internal or external, can significantly impact the bedrock incision. The purpose of this study is to experimentally investigate the bedrock incision under sediment bedload transport along curved channels by varying flow speed with the other parameters kept constant. In this study, a bedrock incision was simulated in an annular flume. Two cases were considered, each with different rotation speeds of the cover lid, using plaster as the bedrock and sediment incision tools. In both cases, sediment motion was the bedload transport. It was found that the sediment deposited along the inner wall differed according to the rotation speed. A uniform transverse slope was found for a rotation speed of 48 RPM and moving bedforms were found for a rotation speed of 40 RPM. The bedrock incision resulted in the formation of the inner channel along the centerline of the flume, which grew more quickly toward the inner wall under the bedforms than under the uniform transverse slope. These findings suggest that the type of bedform has a more significant impact on bedrock incisions than rotation speed.

Keywords: incision; meandering; bedrock; curved channel; bed configurations; annular flume; cover lid rotation speed

Citation: Andriamboavonjy, M.R.; Terakado, T.; Izumi, N. Spatiotemporal Evolution of Bed Configurations in Mixed Bedrock-Alluvial in Uniformly Curved Channels. *Water* **2022**, *14*, 397. https://doi.org/10.3390/w14030397

Academic Editor: Vlassios Hrissanthou

Received: 7 December 2021
Accepted: 23 January 2022
Published: 28 January 2022

1. Introduction

Natural mechanisms for river path changes include alluvial and incision meandering. The former is mainly the result of bank erosion on an alluvial plain, while the latter occurs when a bedrock incision occurs [1]. The floodplains of meandering rivers are usually well-vegetated and rich in fine sediment. The combination of vegetation and fine sediment slows down the erosion and the lateral expansion of the river [2], with a positive impact on the aquaculture of the river system [3]. However, discussions on river incisions tend to be limited to their contribution to the landscape evolution [4,5]. Bedrock incisions were first discussed from an engineering perspective at the end of the nineteenth century, with arguments about their dynamics at the time [6,7]. Incision is commonly active on the bedrock, and it plays an essential role in landscape evolution.

The main parameters influencing bedrock incisions, according to previous studies, are sediment distribution, channel width, and flow fluctuation [8]. Sklar and Dietrich [9] suggested that bedrock incisions occur at the optimal bed coverage and shear stress. On the other hand, Zhang et al. [10] proposed a model considering the local bed state by integrating the bed cover fraction and the ratio of alluvial thickness to bedrock macro-roughness. These two models assume that the bedload transport of the sediment and the sediment grain saltation on the bedrock is the origin of the incision. While these interpretations are widely accepted, few studies have examined the factors that influence bedrock incisions in curved channels. The generation of spiral flow, as seen in Figure 1, and the redistribution of bed shear stress are the main features of flow dynamics in curved channels (e.g., [11]). They are found to impact the shape of the bed topography pattern in a movable-bed channel [12–14].

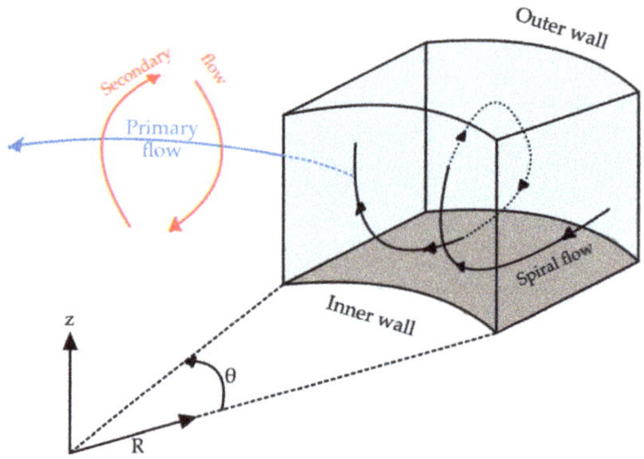

Figure 1. Schematic diagram of the spiral flow along curved channels as the combination of the primary and the secondary flow, which corresponds to tangential and radial flow in an annular flume.

In earlier studies on the effect on the hydrodynamics of curved channels, an annular flume has been widely adopted in laboratory experiments. An annular flume has been shown to allow a continuous spiral flow to be generated similar to that in curved channels during the necessary timespan (e.g., [15–19]). Also, to control the spiral flow, the rotation of the cover lid of the annular flume is varied, and the other parameters are maintained at constant values (water depth and sediment amount, and caliber), e.g., [10,16,17].

In an annular flume, the flow velocity components (tangential and radial) and the bed shear stress increase with the top lid rotation speed (e.g., [16,18]). The tangential flow velocity increases from the inner wall towards the outer wall. Simultaneously, this spiral flow drags the sediment towards the inner bank and forms the transverse slope (e.g., [20–23]). The radial flow velocity has been shown to vary by between 20 and 50% of the tangential flow velocity near the bed [18]. When the spiral flow varies in space and time, the bed topography also fluctuates (e.g., [16,24–26]). It has also been reported that any variation of the channel morphology, bed state, and flow dynamics significantly affect the spiral flow characteristics (e.g., [10,16]). On the other hand, it has also been reported that the bed shear stress rises from the inner wall to the outer wall, except at the regions near the walls: right after the beginning of the cover lid rotation, the minimal second flow criteria are met, and the sediment is deposited along the inner wall, forming a uniform transverse slope which gradually turns into regular deposition patterns (e.g., [12,16,24,26]). Taguchi et al. [27,28] discovered a limited incised area experimentally and concluded that the erosion mainly occurs under the moderately covered bed, namely at the base of the transverse slope. They also suggested the alternance of the covered and exposed bed status as one more reason for the erosion. Their results were obtained with a constant cover lid rotation speed of 40 RPM, and mortar simulated the bedrock. It should be noted, however, that the erosion reported by Taguchi et al. [28] was likely only a small part of the erosion inside the annular flume. No analysis of the advanced state of the erosions that they found was included in their study.

In this investigation, we explore bedrock incisions experimentally along uniformly curved channels in the annular flume in a manner similar to that used by Taguchi et al. [27,28]. In our study, however, we use the plaster model of the bedrock for simulation purposes and we use different rotation speeds of the cover lid. The objectives of this investigation are to determine the precise locations of the occurrence of incisions along the uniformly curved channel, and the mechanism of the erosion in terms of space and in time and how that affects the sediment deposition patterns.

2. Materials and Methods

We conducted two experiments with different cover lid rotation speeds: 40 RPM and 48 RPM, which will be referred to as Case 1 and Case 2, respectively, in this paper.

2.1. Annular Flume

We conducted the experiments in an annular flume made of plexiglass, as seen in Figure 2. Figure 3 shows the schematic diagram of the annular flume. It has a rectangular cross-section with a width B = 100 mm, and a maximum depth Hmax = 180 mm, and a radius at the centerline of Rc = 450 mm. We adjusted the cover lid to have h = 50 mm of the water column. The rotation of the cover lid, placed on the water column, permitted the flow to generate inside the flume.

Figure 2. Photo of the annular flume with the white plaster. The flume is made of plexiglass, the cover lid of metal for better stability, and the plate in contact on the water surface is also made of plexiglass.

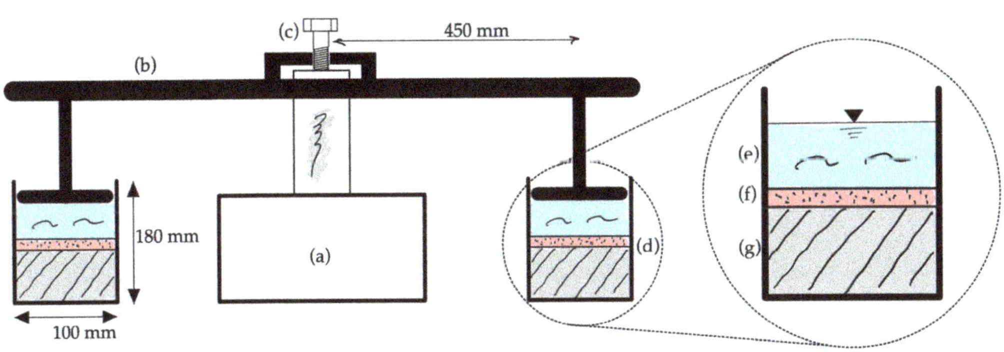

Figure 3. Schematic diagram of the cross section of the annular flume and the original bed configuration inside the flume channel. (**a**) electrical motor, (**b**) cover lid, (**c**) lever for the cover lid plate to obtain the desired water depth, (**d**) flume channel, (**e**) 50 mm of water depth, (**f**) 2.5 kg of sediment, (**g**) 6 cm of plaster, a mixture of gypsum powder and water.

2.2. Bed Materials

We used plaster, a mixture of gypsum and water, as bedrock. We used the HS-650® gypsum provided by Noritake Co., Ltd., Nagoya, Japan. [29]. After the mixing process, the plaster was poured into the flume and was left for 12 h, then the excess water was pumped out, and the plaster was left to cure for 24 h. The amount of plaster used in our studies was such that it reached a thickness d = 60 mm. Plaster was used because it is easier and faster to prepare than mortar [27,28,30,31]. In addition, plaster has a lower resistance against erosion [32]. For Case 1 and Case 2, the plaster mixtures were 1:2 and 1:2.5 by weight of gypsum and water, respectively. It means that the bedrock in Case 2 was slightly softer than in Case 1. Regardless, the bedrock preparation was identical for both cases.

As sediment, 2.5 kg of sand was used. It is a natural sand (No. 5® provided by Tohoku Keisya Co., Ltd.) and it has a diameter D_{50} = 0.45 mm with a density of ρ_s = 2.61 g/cm^3. Its grain-size distribution curve is presented in Figure A1, Appendix A [33]. The sand was painted red to contrast the bedrock with the sediment [27,28]. The sediment was dispersed on the bed surface before each start, as uniformly as possible. Water was then poured carefully into the flume to avoid the possibility of sediment disruption and any damage to the bed surface due to the water jet. For these reasons, the pouring flow was kept as low as possible, and the water was passed through a small-screen sponge. In our experiments, the water depth was 50 mm, and any excess water was removed for safety reasons.

2.3. Experimental Conditions and Data Acquisition

As previously stated, two experimental cases were conducted. In Case 1, the rotation speed was 40 RPM of top lid rotation, and in Case 2, it was 48 RPM. The studies by Taguchi et al. [27,28] were used as a guide for the speed in Case 1, and in Case 2 the aim was simply to investigate the higher rotation speed effects on the erosion. The amount of sediment remained constant throughout the experiments, and the sediment transport was kept to bedload transport, which was checked visually at the start of each experiment.

We ran the experiments within specified timesteps (Table 1). After each timestep, the bed surface topography was surveyed twice: first, the bed surface with the sediment coverage (WS) was surveyed, and then the bed surface without the sediment coverage (NS) was done.

Table 1. Experimental conditions and resultant bed formation.

	Timestep	Δt [1]	T [2]	Speed [3]	α [4]	W [5]	Bedforms, Number of Wavelengths (k)
Case 1	1	1/12	0	40	23–24	26	Uniform transform slope
	2	3	3	40	8–37	41	5 bedforms
	3	3	6	40	16–28	34	5 bedforms
	4	6	12	40	15–38	37	6 bedforms
	5	12	24	40	15–61	24	8 bedforms
	6	24	48	40	18–55	33	9 bedforms
Case 2	1	1/12	0	48	20–30	20	Uniform transform slope
	2	3	3	48	12–33	24	12 bedforms
	3	3	6	48	15–31	22	12 bedforms
	4	6	12	48	21–26	17	Uniform transform slope
	5	12	24	48	20–26	17	Uniform transform slope
	6	24	48	48	20–22	17	Uniform transform slope
	7	24	72	48	20–27	18	Uniform transform slope
	8	12	84	40	12–38	34	6 bedforms

[1] Δt: the timestep's duration, in hours (h); [2] T: the cumulative time, in hours (h); [3] Speed: the annular flume top lid rotation speed, in rounds per minute (RPM); [4] α: the transverse slope angle, in degrees (°); [5] W: the transverse slope base sweep width, in millimeters (mm).

We used laser scanning to measure the topography of the bed surface. However, the scanning equipment was different for the two cases: Case 1 used a Keyence CMOS Multi-Function Analog Laser Sensor, whereas Case 2 used an Artec EVA 3D scanner. The Sensor head IL-300 was utilized with the Keyence CMOS Multi-Function Analog Laser Sensor and was placed above the bed surface at a distance of 300 mm to provide the optimal precision of 0.5 mm [34]. Similar to the studies of Taguchi et al. [27,28], the bed surface was surveyed circumferentially at every 50 mm from the inner wall side. The 3D scanner Artec EVA of Artec 3D was operated at a 500–600 mm distance from the bed surface to provide better resolution. It has a resolution of up to 0.2 mm [35]. Contrary to the Keyence IL 300

Laser Sensor, the bed surface and the flume were surveyed and provided complete and continuous topography data.

2.4. Data Processing and Calculation

The measured data were in (x, y, z) format, representing the relative bed elevation of the bed surface in the cartesian coordinates system. It is then changed into a curvilinear coordinates system (s, n, z) for visual convenience. The tangential and radial coordinates (s, n) are respectively defined by the following:

$$n = R - Rc, \tag{1}$$

$$s = Rc \times \theta \tag{2}$$

where Rc is the radius at the centerline of the flume, which is 450 mm, $R = \sqrt{x^2 + y^2}$ is the radius, and $\theta = \tan^{-1}(y/x)$. Hence, s = [0, 900] π mm; and n = [−50, 50] mm. The negative values of n indicate the half side near the inner wall of the annular flume. In contrast, the positive ones represent the half near the outer wall, and 0 is the flume centerline. Note that −50 is located on the inner wall side, and 50 is located on the outer wall.

2.4.1. Transverse Slope

We generated cross-section profiles for each timestep from the bed surface topographical map. We measured the width and the depth of the inner channel along those sections, denoted as Be and He, respectively (Figure 4). Photos of the bed surface are presented in Figure 5. We also retrieved the profile data from the 18 transverse sections equidistant by 50 π mm (Figure 6). The average erosion rate (Em) is obtained by dividing the difference between the sequential bed surface and the original bed surface by the duration.

2.4.2. Sweep Width

The sweep width, denoted as W, was defined as the bedrock surface area inside which the transverse slope base undulates (Figure 4). This sweep width is also retrieved from these cross-section profiles. It is defined by the difference between the minimum and maximum radial coordinates of the transverse slope base on the bedrock surface.

2.4.3. Fraction of Exposed Bed

The fraction of the exposed bed, denoted as Po, was calculated by analyzing the topographical map of the bed surface. First, we generated the sediment thickness maps. Then the spatial analysis tools implemented in ArcGIS [36] were used to delineate and calculate the areas of the exposed bed (Ao) and the sediment coverage (Ac) (Figure 4). To differentiate the sediment coverage to the exposed bed, we applied a criteria value of 0.5 mm on the sediment thickness map, which is the diameter of the grain. Values below 0.5 mm were considered to represent the exposed bed, and higher values were considered sediment coverage. The following formula can be used to describe Po, according to the study of Sklar and Dietrich [9]:

$$Po = Ao/(Ao + Ac) \tag{3}$$

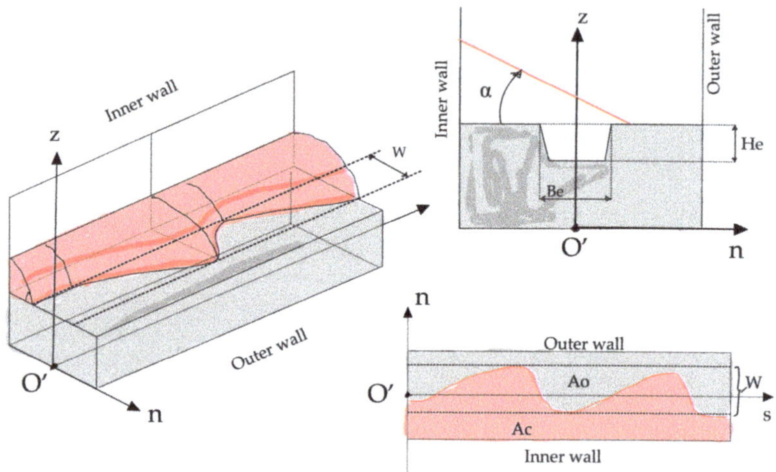

Figure 4. Schematic diagrams of the bed configurations and parameters definition. The sediment coverage is colored in red; the bedrock is gray (**left**). The corresponding plan and section views are on the upper right and bottom right, respectively.

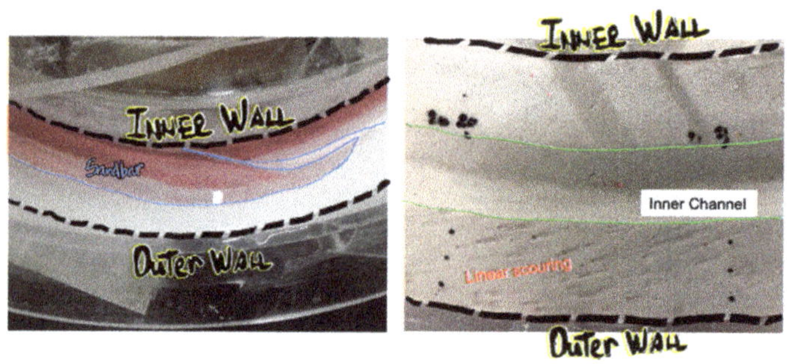

Figure 5. Photos of the bed configuration inside the flume—top view—Case 1 at the end of timestep 6. The white-colored area represents the exposed bed, the reddish part is the sediment coverage (**left**). The inner channel is visible after removing the sediment coverage, and linear scouring is visible under magnification (**right**).

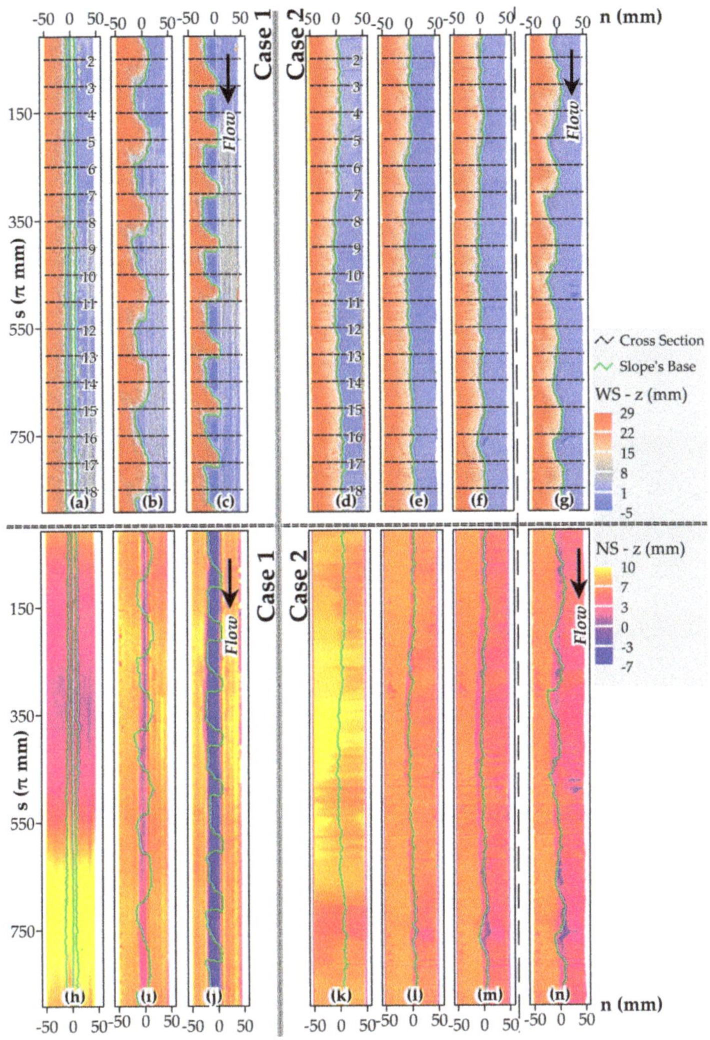

Figure 6. Plan views of the bed configurations at the timesteps 1 (T = 5 min), 4 (T = 12 h), and 6 (T = 48 h). The horizontal and vertical axes represent (n, s) the radial and tangential coordinates, respectively. The radial coordinates n = −50 and 50 indicate the inner and outer wall sides, respectively. (**a–c**), and (**d–f**) represent the sediment deposition patterns in Case 1 and Case 2, respectively. (**h–j**), and (**k–m**) represent the evolution of the inner channel, the bedrock incision result in Case 1 and Case 2, respectively. (**g,n**) represent the bed configurations at the timestep 8 in Case 2, after lowering the rotation speed to 40 RPM. The green curves indicate the boundary between the exposed bed and the sediment coverage, and the dashed lines represent the transverse sections for the profiles data for all 18 cross-sections.

3. Results

The sediment deposition patterns with the corresponding transverse slope profiles and the evolution of the bedrock incision were determined for Case 1 and Case 2. Figure 5 shows two photos of the resultant bed configurations for Case 1. The sediment bed coverage constituting the bedforms (red-colored) was deposited along the inner wall, and the exposed bed (white-colored) along the vicinity of the outer wall could be observed

(Figure 5, left). The bed incision resulted in the formation of an inner channel along the centerline of the flume. Moreover, light and linear scours were also noticed on the exposed bedrock (Figure 5, right).

Figure 6 shows the plan views of the bed topography for both cases, the sediment deposition patterns, and the bedrock surface erosion. The WS represents the bed surface with the sediment (Figure 6a–g), whereas NS indicates the bed surface after removing the sediment coverage. (Figure 6h–n) The timesteps 1, 4, and 6 are displayed to show the temporal evolution of the bed configurations. Timestep 8 of Case 2 is also presented here, indicating the result after slowing the rotation speed to 40 RPM (Figure 6g for WS and Figure 6h for NS). The plan view over the whole series of timesteps can be found in Appendices B and C (Figures A2–A5).

3.1. Sediment Deposition Patterns

The sediment transport inside the flume started soon after the experiment began. Initially, the sediment was dragged toward the inner wall due to the spiral flow and formed a uniform transverse slope along the inner wall. The uniform transverse slope was observed in Case 1 (Figure 6a), as well as in Case 2 (Figure 6d). It was found that the sediment generally covered half the flume width.

Gradually, the uniform transverse slope turned into migrating bedforms, depending on the in-flume conditions. For instance, the bedforms were observed throughout Case 1, their wavenumber increased, and their shape changed with time (Figure 6b,c). In Case 2, there was little evidence of bedforms and the sediment deposition along the inner wall side looked like a uniform transverse slope (Figure 6e,f). At the timesteps 2 and 3, twelve bedforms were observed (Table 1). There were also six bedforms at timestep 8, when the rotation speed was lowered to 40 RPM (Figure 6g).

3.2. Transverse Slopes and Sweep Width

Figure 7 shows the transverse slope profiles of the sediment coverage, which were presented in Figure 6. The transverse slope profiles over the whole series of timesteps can be found in Appendix D (Figures A6 and A7). We measured the transverse slope angle (α) and the transverse slope base sweep width (W). These are reported in Table 1.

For both cases, the average transverse slope is similar, at between 23 and 24 degrees (Figure 7). In Case 2, less variation was found for the transverse slope angle α, between 20 and 27 degrees, except at timesteps 2, 3, and 8, where the fluctuation of the angle α became significant, namely between 12 and 38 degrees (Figure A7b,c,h). In contrast, in Case 1, the variation of the angle α was significant, from 8 to 61 degrees (Figure 7, right and Figure A6.

In addition, with the formation of the bedforms, the sweep width (W) became wider. As seen in Figure 5, Case 1 showed well-formed bedforms (long wavelength), and W became wider than 25 mm (Table 1). In contrast, when the bedforms were poorly formed (short wavelength) or remained as a quasi-uniform transverse slope as in Case 2, W became narrower, at less than 25 mm (Figure 6d–f, Table 1).

3.3. Fraction of Exposed Bed (Po) and Average Erosion Rate (Em)

Figure 8a illustrates the plots of the fraction of exposed bed (Po) and the average erosion rate (Em) over time, revealing a close link between the two. A rise in Po was accompanied by a rise in Em.

In both Case 1 and Case 2, around 45% of the bed surface is exposed at first. Then, until T = 6 h, Em and Po grew and achieved local maxima. Following that, Em and Po in Case 1 increased, reaching 1.78×10^3 mm^3/h and 56%, respectively, by the end of the experiment T = 48 h. In contrast, in Case 2, Em and Po decreased and stabilized at 4.8×10^3 mm^3/h and 47%, respectively. Particularly in Case 2, Po increased and reached 52%, leading to a slight increase of Em by the end of the experiment when the rotation speed was reduced to 40 RPM.

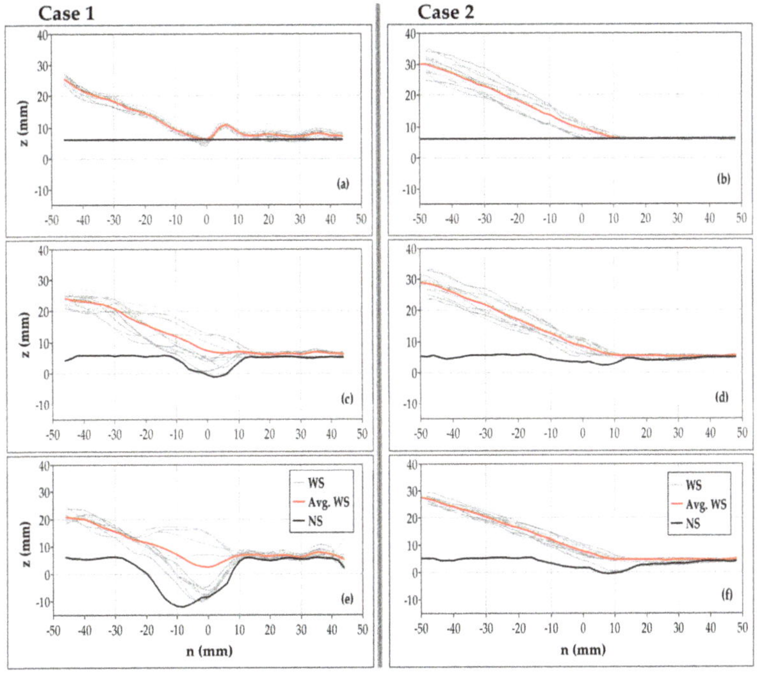

Figure 7. Transverse profile sections—Case 1 (**left**) and Case 2 (**right**). The horizontal and vertical axes represent the radial and vertical coordinates, respectively. The average of the bedrock surface profiles (NS, black color), the 18 profiles of the sediment coverage (WS, gray color), and their average (Avg. WS, red color) are indicated from top to bottom, at timesteps 1 (T = 0 h), 4 (T = 6 h), and 6 (T = 48 h) (**a**,**c**,**e**) for Case 1, and (**b**,**d**,**f**) for Case 2).

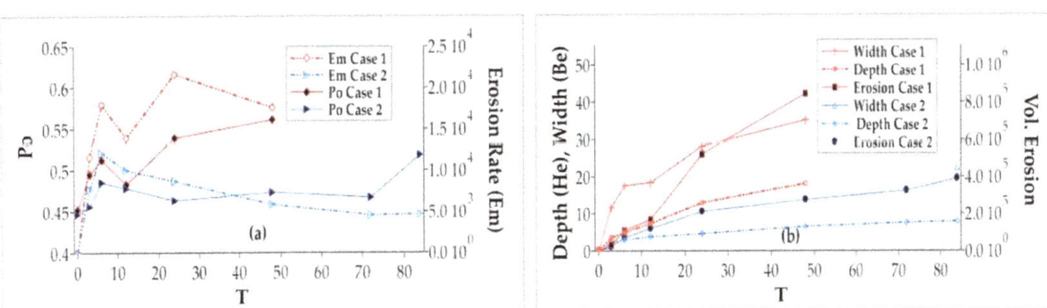

Figure 8. (**a**) Graphs of the mean erosion rate (Em, dashed lines, mm^3/h) and the fraction of exposed bed (Po, solid line) for Case 1 (in red) and Case 2 (in blue). The horizontal axis is the cumulative time T (h); the vertical axes represent Po on the left and Em on the right; (**b**) Temporal evolution of the erosion for Case 1 (red) and Case 2 (blue) over cumulative time T (h). The vertical axis on the left represents the depth (He) and the width (Be) of the inner channel (mm), and the right axis represents the volume of erosion (mm^3/h).

3.4. Bedrock Erosion

As was explained, the water and sediment depth in Case 1 and Case 2 was identical, and the only difference was the cover lid rotation speed, at 40 RPM in Case 1 and 48 RPM in

Case 2. This made it possible to investigate the effect of rotation speed on abrasion erosion on the bedrock surface by comparing the two cases.

First, the erosion in Case 2 was less than half that of Case 1. In fact, at the end of timestep 6, the total volume of erosion in Case 1 was 800,000 mm^3, whereas it was only 300,000 mm^3 in Case 2 (Figure 8b). In general, the erosion graphs grew quickly from the start until T = 6 h, and gradually increased afterward. However, a sudden rise in the erosion appeared at T = 24 h in Case 1, and at T = 84 h in Case 2.

Second, the lateral expansion of the inner channel was faster than the vertical expansion. The increase in width (Be) was more significant than the increase in depth (He) throughout the experiments in both cases. At the end of the experiments, Be was 35 and 22 mm in Cases 1 and 2, respectively, while He was 18 and 8 mm, respectively (Figure 8b).

The plan views of the bedrock surfaces are displayed in Figure 6, and the corresponding profile sections are presented in Figure 7 as NS. Growth in the inner channel, considered primary erosion, was observed along the boundary between the bed coverage and the exposed bed, inside the sweep width (W) area. This boundary is indicated by the light-green curve in Figure 6.

Finally, Case 1 had clearly obtained a much more advanced stage of inner channel compared to Case 2. The bedrock was vertically incised first, then downcut toward the inner wall (Figure 9a). The vertical incision was observed until T = 12 h in Case 1, while in Case 2 (Figure 9b), it was observed until T = 84 h. In Case 1, the incision toward the inner wall was clearly visible from T = 12 h.

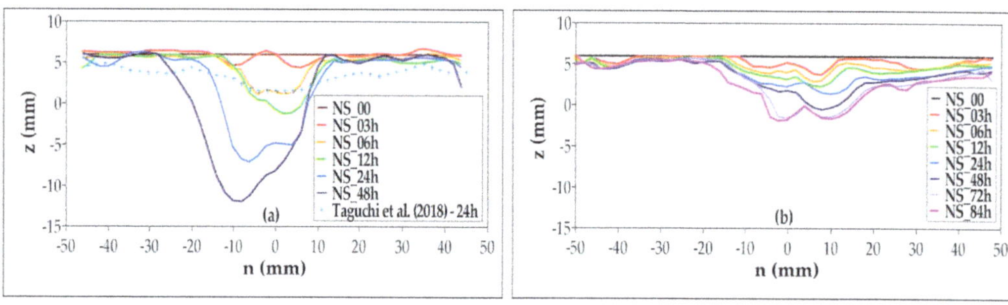

Figure 9. Circumferentially averaged bedrock surface profiles evolution with the cumulative time T, Case 1 (**a**) and Case 2 (**b**). The radial and vertical coordinates (n, z) are represented by the horizontal and vertical axes, as shown in Figure 7. The centerline of the flume is represented by n = 0, with negative values on the half side near the inner wall and positive values on the outer wall side. The measured mortar bedrock surface level by Taguchi et al. [28] at T = 24 h are plotted as blue crosses.

4. Discussion

The results reveal the mechanism involved at the onset of the erosion. Also, the spatiotemporal evolution of the inner channel and the sediment deposition pattern found during the experiments allow us to determine how erosion evolves in space and time.

4.1. The Location of Erosion

The formation of the inner channel was along the boundary between the bed coverage with sediment and the exposed bedrock. This boundary is located around the channel centerline. Furthermore, if looking into the transverse slope, this boundary was at the base of the transverse slope, where the sediment coverage is relatively thin. A close observation of the incision of the bedrock surface at the end of timestep 2 reveals that the bedrock was downcut vertically (Figure 7).

While it was not possible to accurately determine the thickness of sediment coverage under which the bed incision occurs, we could graphically estimate that it required about 10 mm in the experimental conditions employed in this study, or approximately 20 times the sediment grain size. It should be noted, however, that this estimation was deduced from the transverse slope at the end of each timestep, meaning that it was at its angle of repose. This suggests that the thickness of the deposit was less than 10 mm in our study.

Under that thickness, the sediment grains can saltate, hit, and wear the bedrock surface. When the coverage exceeds that boundary, no erosion occurs because the motion of the sediment grains cannot reach the bed surface and can be considered a fully covered bed. According to previous researchers, that boundary is the place of "tradeoff between the availability of abrasive tools and the partial alluviation of the bedrock" [9] or the toe of the alluvial bedforms [12] or the portion moderately covered with sediment [27,28]. In sum, the base of the transverse slope would be the only place where the sediment grains may collide with the bedrock surface [30].

Increased cover lid rotation speed, on the other hand, had no effect on location of erosion occurrence. That is, the boundary between the sediment coverage and the exposed bed was positioned at the same place across the flume width on both Case 1 and Case 2. This can likely be attributed to the mean transverse slope and the amount of sediment used, since they specify the position of the boundary between the sediment coverage and the exposed bed, as well as the location of the incision. This is consistent with Engelund's findings [21] that the mean transverse slope was almost independent of the top lid rotational speed as long as the sediment grains flowed as bedload. Taguchi et al. [28] also obtained similar results to the current study with the same amount and caliber of sediment. From these results it can be concluded that the border between the sediment-covered bed and the exposed bedrock where erosion occurs is essentially the same regardless of how fast the cover lid rotates.

We also observed linear scours on the exposed bed surface even though erosion should not have happened on the exposed bed based on the abrasion models. These scours, we believe, were the consequence of interaction between sediment grains and flow directly on the bedrock surface, with the fine structure of the gypsum powder in the plaster allowing this interaction engraved on the bed surface (Figure 5, right). If this is the case, these linear scours should be investigated further for a greater understanding of flow dynamics at the bedrock-fluid interface.

4.2. Spatiotemporal Evolution of the Bedrock Incision and the Inner Channel

4.2.1. The Bedrock Incision

It was found that the rate and spatial extension of the bedrock incision are inversely proportional to the cover lid rotation speed. That is, a faster rotation speed is associated with less erosion. This is due to the sediment deposition patterns imposed by the flow conditions. With a higher rotation speed (48 RPM in Case 2), uniform transverse slope formed, and the incision occurred only along its base. With a slower rotation speed (40 RPM in Case 1), migrating bedforms formed, leading to an increase in the bed area subjected to erosion, which can be attributed to the undulations of the bedform boundary (Figures 6 and 7). Also, when reducing the rotation speed from 48 to 40 RPM (Case 2, timestep 8), the bedforms reappeared (Figure 6g), resulting in a greater erosion rate at the end of the experiment in Case 2 (Figure 8a).

The bedforms vary with rotation speed because the speed influences the secondary flow (e.g., [16,24]), and with time as the bedforms fluctuate and stabilize at a "mature" stage (e.g., [21,24,26]).

In addition, when the undulating bedforms are formed, their migration allows for the alternation of "covered" and "exposed" bedrock surfaces, a phenomenon known as runaway alluviation [37]. This runaway alluviation combined with the saltation of the sediment grains and the undulation of the boundary explains the higher erosion in Case 1 than in Case 2. In the latter, no runaway alluviation was observed. Our results provided one quantitative example of runaway alluviation to add to the results obtained in previous studies (e.g., [8,20,21,31]).

4.2.2. The Inner Channel

Initially, the incision vertically downcut the bedrock, slightly shifting toward the outer wall (Figure 9). This vertical incision occurred at the center of the sweep width (W), or right under the boundary of sediment coverage and the exposed bed, because this location had a high probability and high frequency of erosion. This high probability of erosion was due to the primary conditions of incision and high frequency due to its position inside the sweep width.

The advanced state of the incision shows that it gradually downcut towards the inner wall (Figure 9a). This shift resulted from the alternating lateral shift toward the inner wall and the downward incision on the bedrock surface. Because no extra sediment was supplied, the bedrock near the outside wall was left exposed. Because the sediment was only available along the inner wall side due to the spiral flow, it is evident that the conditions for maximum incision also move towards the inner wall. As a matter of fact, the height of the bedform gradually decreased with the growth of the inner wall, suggesting that the incision would erode the bedrock along the inner channel at some point.

This incision toward the inner wall is consistent with the results reported by Fernández et al. [12] using a meandering channel of highly fixed sinuosity in the absence of sediment supply. Shepherd [31] also discovered similar findings in his experiment along the manually excavated sinuous channel and reported that the erosion went toward the inside of the bend until the gradient and the velocity dropped, allowing the sediment to deposit, and thereafter the erosion process inverted toward the outer wall.

However, this was not reported by Taguchi et al. [27,28] in their experiment. This may be explained by the geometry of the inner channel. In this study, the simulated bedrock, made of plaster, provided faster erosion. For instance, at T = 24 h, the inner channel size was larger in plaster than in mortar (Figure 9a). The flow dynamics combined with this growth of the inner channel and the sediment transport inside the flume likely explain the actual result.

In our study, there was no sediment supply, and the rotation speed was maintained at a constant so that the erosion would continue toward the inner wall. However, further investigation is required to determine the progress of the inner channel.

4.3. Deposition Pattern

Along uniformly curved channels, the bedforms depend on the rotation speed, time, and development of the inner channel resulting from the bedrock incision.

First, the averaged transverse slope angle for both cases was similar, at about 23 degrees, throughout the timesteps (Figure 7). This averaged angle represents the angle of the mean bedforms and was found to be independent of the rotation speed in case of sediment bedload transport [24]. However, the difference in the slope angles in Case 1 and Case 2 may be related to the timing of the measurements and the rotation speed. The slightly higher transverse slope angle and height in Case 2 than those in Case 1 can likely be attributed to the faster rotation speed of Case 2 [24]. However, we should keep in mind that because these measurements were made after the experiments were stopped, the obtained values may be underestimated.

Second, while the secondary flow increases when the rotation speed is high, the spiral flow wavelength, as well as the wavelength of the bedforms decreases. Case 2, at 48 RPM, for example, had poorly formed bedforms that were hardly discernible (Figure 6). When the rotation speed was increased, Engelund [21] discovered a similar result, and attributed it to the reformation of the uniform transverse slope. Inversely, with a slower rotation speed, the secondary flow decreases, and the effects of the spiral flow on the bedforms could be well observed, as we could witness the well-formed bedforms in Case 1 at 40 RPM. After timestep 2, for example, there were 5 and 12 bedforms in Case 1 and Case 2, respectively. This is consistent with the results reported by Baar et al. [24] and Kikkawa et al. [23] when observing the effect of the variation of the secondary flow intensity on the bed morphology.

Third, the number of bedforms increases with the growth of the inner channel. When the rotation was lowered from 48 to 40 RPM, as in Case 2 timestep 8, the bedforms reappeared, with a total of 6 bedforms (Figure 6g). However, for the same 12 h of experimentation in timestep 4 in Case 1, there were 8 bedforms (Table 1). This discrepancy could be attributed to the difference in the dimension of the inner channel, which is more significant in Case 1 than in Case 2 (Figure 9). A larger inner channel was associated with an increase in the number of bedforms from 5 to 9, as seen in Case 1 (Figure 6).

Finally, another feature revealed by the experimental results is the reformation of the uniform transverse slope over a longer experiment run, as indicated in Case 2 (Figure 6). The bedforms gradually faded away as the rotation speed increased, restoring the uniform transverse slope. The gradual convergence of the sediment surface towards the uniform transverse slope can be seen in the section profiles (Figure 7). However, these results do not agree with the results reported by Engelund [21] and Taguchi et al. [27]. They claimed that the number of bedforms first increased and then decreased in number as the experiment progressed. In Engelund's experiment, the erosion was not part of the study, whereas in Taguchi et al.'s experiment, the erosion was not significant compared to the erosion observed. Most likely, the development of the inner channel was the reason for this discrepancy (Case 1). While it remains unclear, it appears that the growth of the inner channel adds to the complexity of the flow dynamics. On the other hand, this discrepancy can also be the outcome of the bedforms splitting and merging processes in Case 2, leading to an equilibrium state at a mature stage of the experiment. This is a topic for further investigation.

5. Conclusions

Erosion by incision of the bedrock along uniformly curved channels was successfully simulated using the annular flume. Our study covered two experimental cases with different rotation speeds but the same sediment and water depth. The rotation speeds significantly impacted the sediment deposition and transport. The bed configurations measurements revealed the formation of the uniform transverse slope and bedforms, and an inner channel formed in the bedrock as the result of the incision. The sediment deposition was found along the inner wall, whereas the bedrock incision occurred around the centerline. Their evolution with time and space was found to be closely dependent.

The findings of this study reveal that while rotation speed does not significantly impact the transverse slope angle, it does condition the formation of bedforms. The initial uniform transverse slope gradually turns into bedforms. At 40 RPM, they are well formed, whereas at 48 RPM, the bedforms are poorly formed. In addition, it is possible that the bedforms would turn back into uniform transverse slope if the experiment had been continued for a longer period of time, which would have allowed the bedforms to split and to merge sufficiently.

The bedrock incision occurs mainly at the base of the transverse slope, and the growth of the inner channel depends on whether the bed deposition pattern constitutes uniform transverse slope or bedforms. The base of the transverse slope or the boundary of the exposed bed and the sediment coverage appear to be linear under the uniform transverse slope and undulating inside the sweep width when the bedforms form. The erosion is minimal when the transverse slope is uniform along the flume, leading to a vertical, shallow, and narrow channel. In contrast, the erosion is significant when there are bedforms: the spatiotemporal variation of the bedforms conditions the development of the inner channel. When there is no sediment supply, the inner channel progresses towards its convex side, leaving the concave side and the bedrock surface near the outer wall of the flume intact. The findings of this study reveal the need to focus our future research efforts on the sediment transport and deposition within the annular flume, and the effects of the inner channel on the flow dynamics and the sediment transport and deposition in an annular flume, as well as the translation of the findings to natural rivers.

Author Contributions: Conceptualization, M.R.A. and N.I.; methodology, M.R.A.; software, M.R.A. and N.I.; investigation, M.R.A. and T.T.; data curation, M.R.A. and T.T.; writing—original draft preparation, M.R.A.; writing—review and editing, M.R.A. and N.I.; visualization, M.R.A. and T.T.; supervision, N.I. All authors have read and agreed to the published version of the manuscript.

Funding: This research received no external funding.

Institutional Review Board Statement: Not applicable.

Informed Consent Statement: Not applicable.

Data Availability Statement: The data presented in this study are available on request from the corresponding author.

Acknowledgments: We would like to thank to the River and Watershed Engineering Laboratory for the experimentation.

Conflicts of Interest: The authors declare that they have no conflicts of interest.

Appendix A

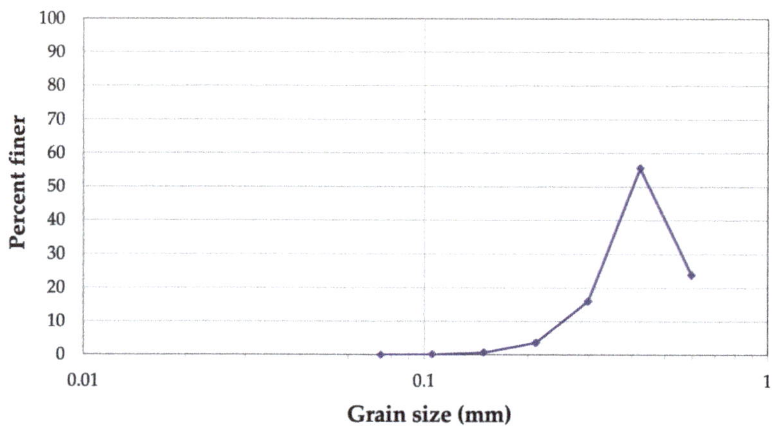

Figure A1. Grain—size distribution curve of the sand No.5® used during the experiments [33].

Appendix B

Plan views of the bed configurations—sediment deposition patterns are presented in this appendix. The horizontal and vertical axes represent (s, n) the tangential and radial, respectively. The radial coordinate n = −50 indicates the inner wall side, whereas n = 50 indicates the outer wall one. The green curves indicate the boundary between the exposed bed and the sediment coverage, and the dashed lines represent the cross-sections for the profile data for all 18 cross-sections.

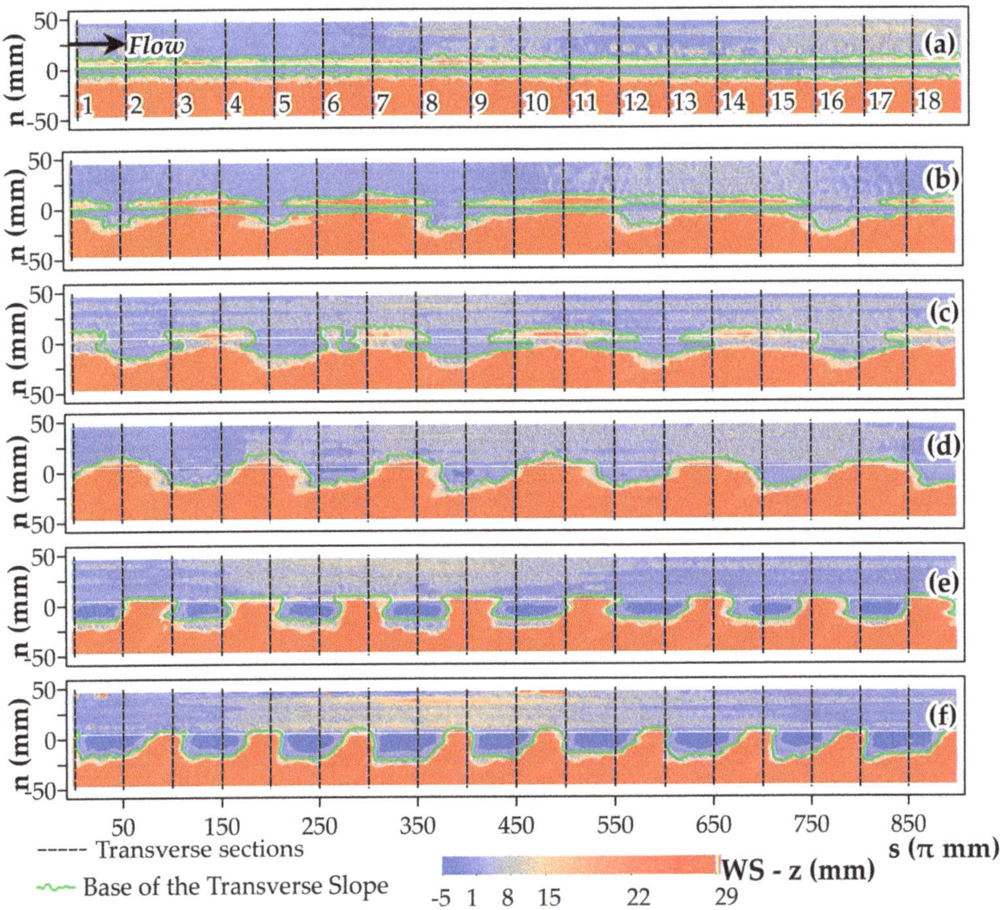

Figure A2. Plan views of the bed configurations—sediment deposition patterns, Case 1 (40 RPM) for all the timesteps: (**a**) timestep 1 (T = 5 min), (**b**) timestep 2 (T = 3 h), (**c**) Timestep 3 (T = 6 h), (**d**) timestep 4 (T = 12 h), (**e**) timestep 5 (T = 24 h), and (**f**) timestep 6 (T = 48 h).

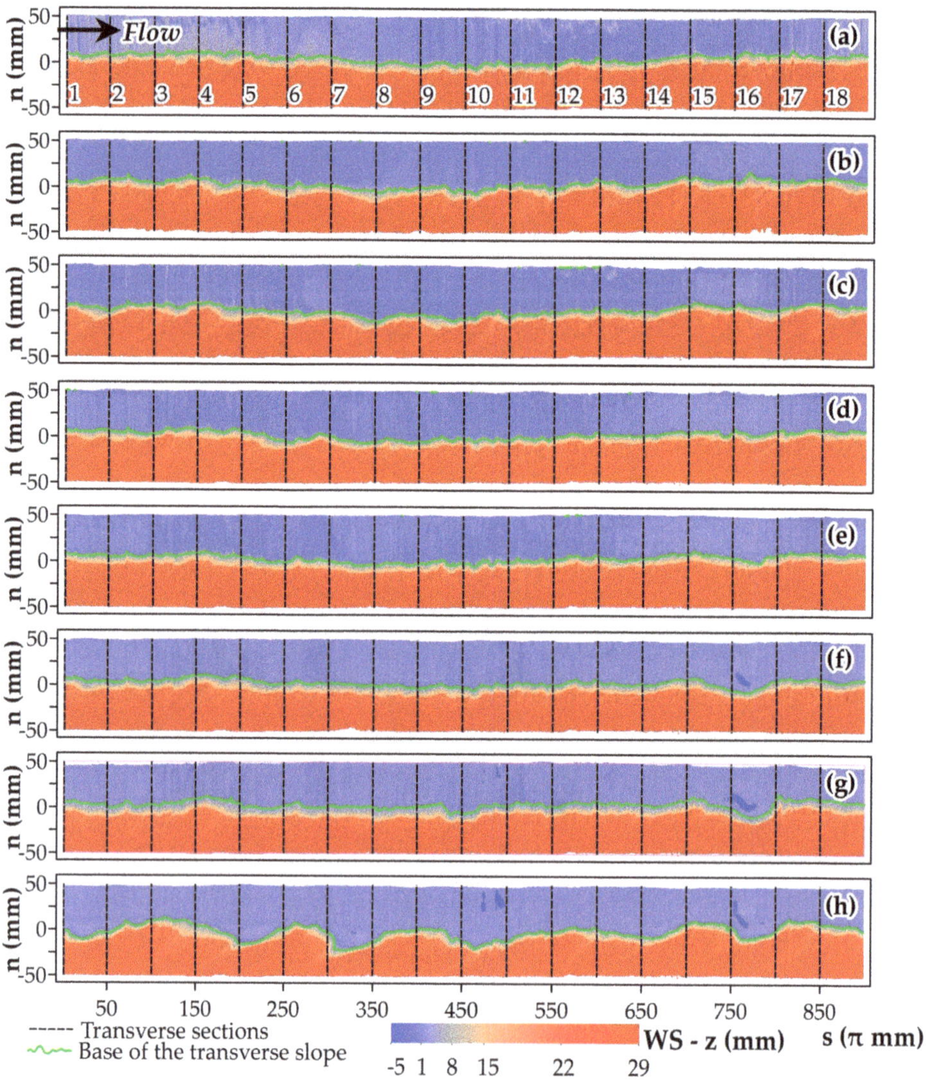

Figure A3. Plan views of the bed configurations—sediment deposition patterns, Case 2 (48 RPM) for all the timesteps: (**a**) timestep 1 (T = 5 min), (**b**) timestep 2 (T = 3 h), (**c**) timestep 3 (T = 6 h), (**d**) timestep 4 (T = 12 h), (**e**) timestep 5 (T = 24 h), (**f**) timestep 6 (T = 48 h), and (**g**) timestep 7 (T = 72 h); (**h**) timestep 8 (T = 84 h) but with 40 RPM.

Appendix C

Plan views of the bed configurations—The bedrock surfaces are displayed in this appendix. The horizontal and vertical axes represent (s, n) the tangential and radial, respectively. The radial coordinate n = −50 indicates the inner wall side, whereas n = 50 indicates the outer wall one. The green curves indicate the boundary between the exposed bed and the sediment coverage.

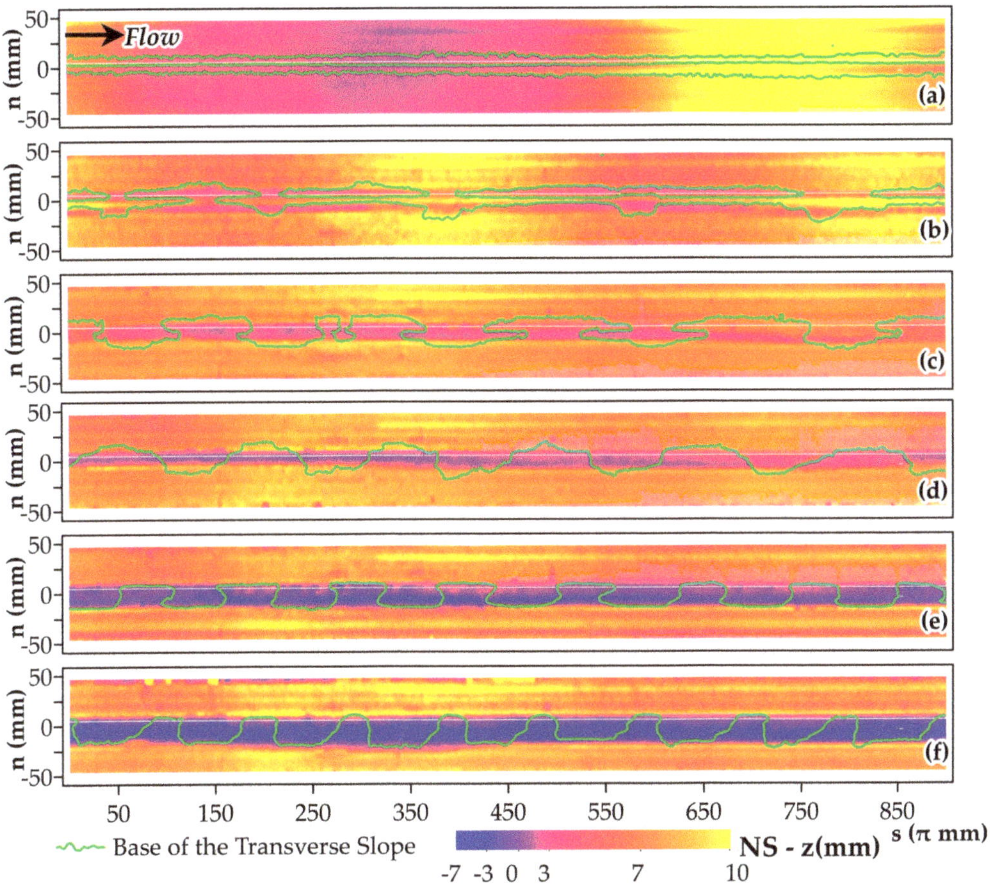

Figure A4. Plan views of the bed configurations—Bedrock surface, Case 1 (40 RPM) for all the timesteps: (**a**) timestep 1 (T = 5 min), (**b**) timestep 2 (T = 3 h), (**c**) timestep 3 (T = 6 h), (**d**) timestep 4 (T = 12 h), (**e**) timestep 5 (T = 24 h), and (**f**) timestep 6 (T = 48 h).

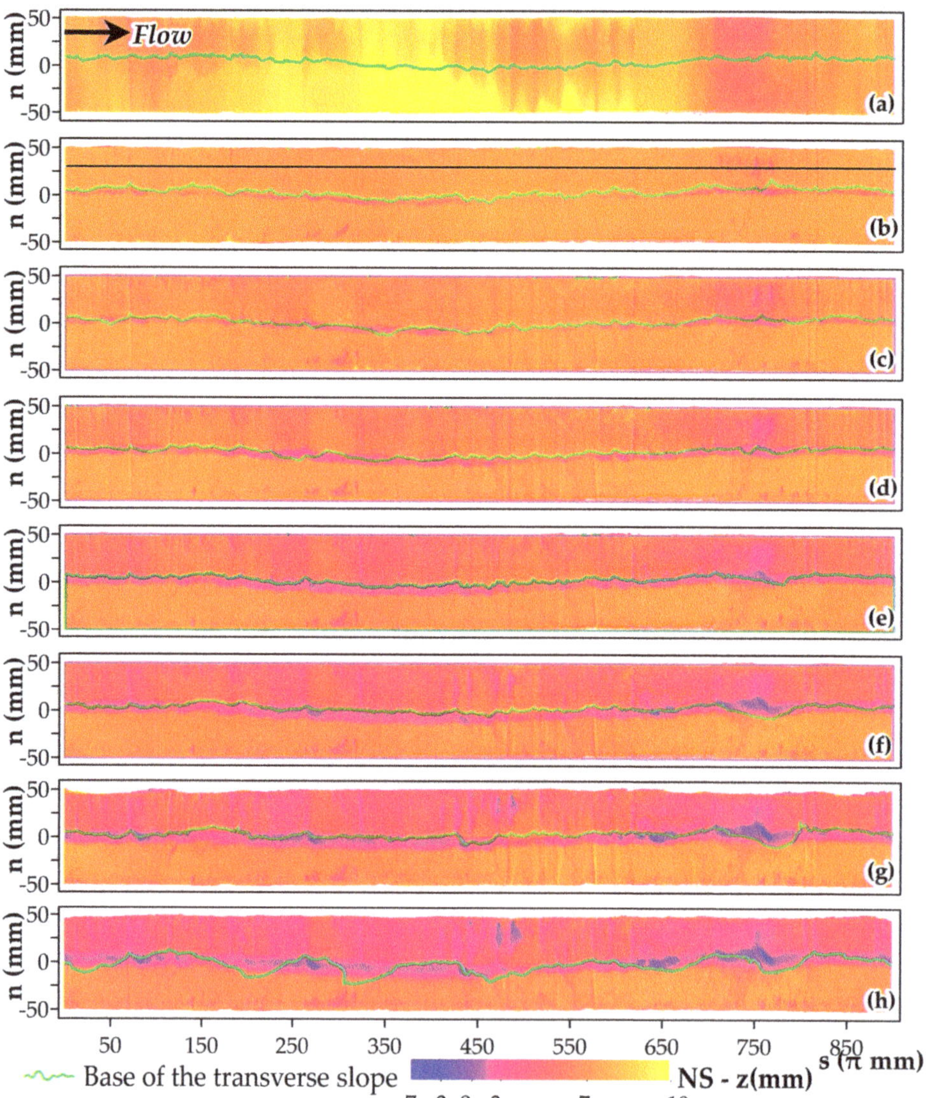

Figure A5. Plan views of the bed configurations—Bedrock surface, Case 2 (48 RPM) for all the timesteps: (**a**) timestep 1 (T = 5 min), (**b**) timestep 2 (T = 3 h), (**c**) timestep 3 (T = 6 h), (**d**) timestep 4 (T = 12 h), (**e**) timestep 5 (T = 24 h), (**f**) timestep 6 (T = 48 h), and (**g**) timestep 7 (T = 72 h); (**h**) timestep 8 (T = 84 h) but with 40 RPM.

Appendix D

Transverse profile sections obtained for each timestep are plotted in this appendix. The horizontal and vertical axes represent (n, z) the radial and vertical coordinates, respectively. The radial coordinate n = −50 indicates the inner wall side, whereas n = 50 indicates the outer wall side. The black curves represent the average profile of the bedrock surface (NS), the red-colored curves are the average of the 18 profiles of the sediment coverage (WS, gray color).

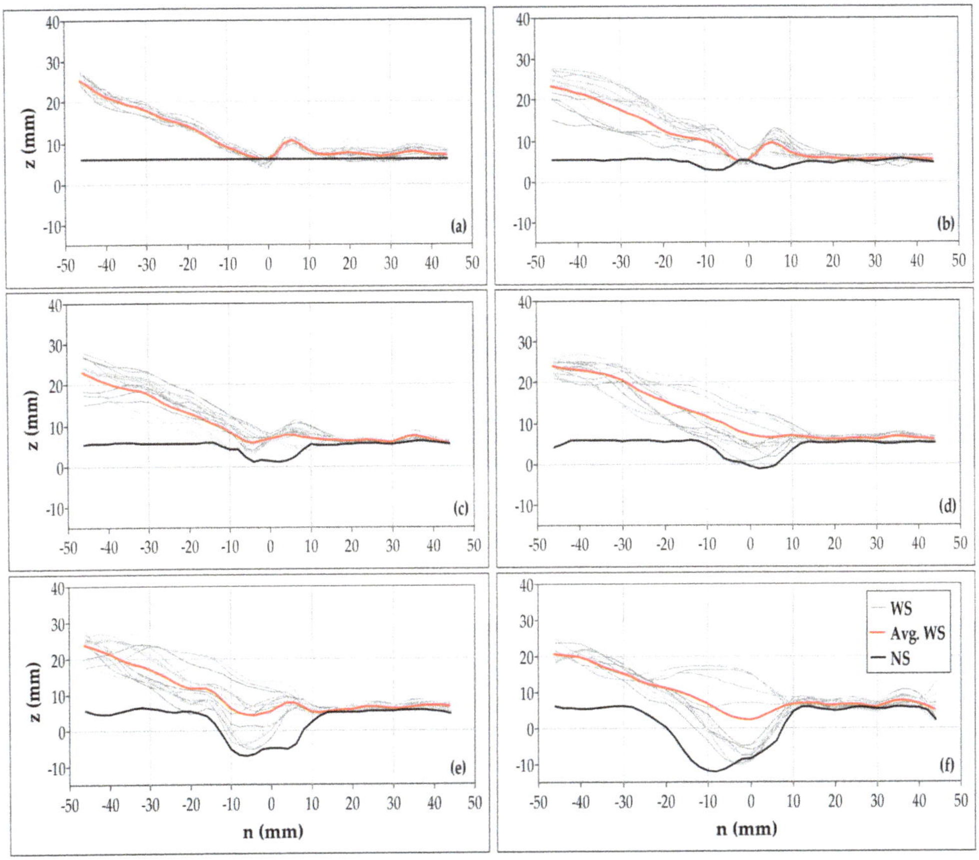

Figure A6. Transverse profile sections obtained for each timestep, Case 1 (40 RPM): (**a**) timestep 1 (T = 5 min), (**b**) timestep 2 (T = 3 h), (**c**) timestep 3 (T = 6 h), (**d**) timestep 4 (T = 12 h), (**e**) timestep 5 (T = 24 h), and (**f**) timestep 6 (T = 48 h).

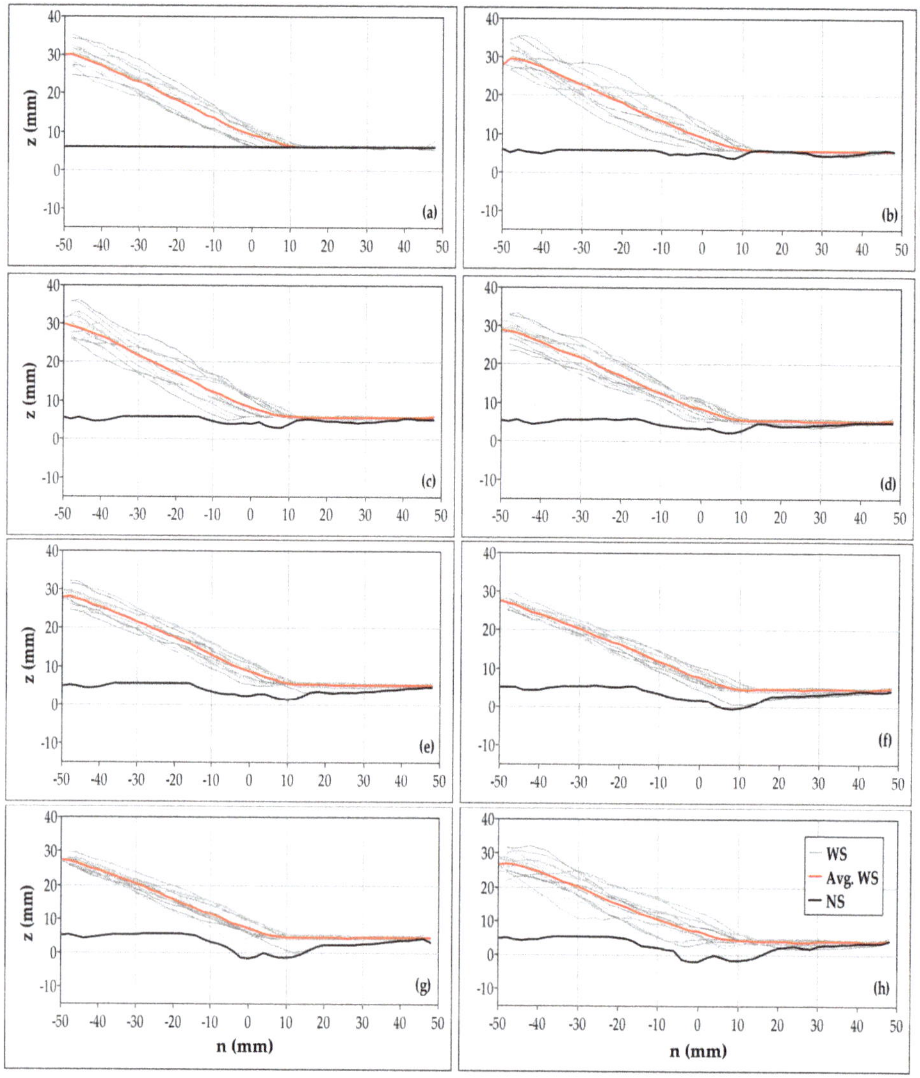

Figure A7. Transverse profile sections obtained for each timestep, Case 2 (48 RPM): (**a**) timestep 1 (T = 5 min), (**b**) timestep 2 (T = 3 h), (**c**) timestep 3 (T = 6 h), (**d**) timestep 4 (T = 12 h), (**e**) timestep 5 (T = 24 h), (**f**) timestep 6 (T = 48 h), and (**g**) timestep 7 (T = 72 h); (**h**) timestep 8 (T = 84 h) but with 40 RPM.

References

1. Kirkby, M. Alluvial and Non-Alluvial Meanders Author(s). *R. Geogr. Soc* **1972**, *4*, 284–288.
2. Parker, G. River meanders in a tray. *Nature* **1998**, *395*, 111–112. [CrossRef]
3. Nakano, D.; Nakamura, F. The significance of meandering channel morphology on the diversity and abundance of macroinvertebrates in a lowland river in Japan. *Aquat. Conserv. Mar. Freshw. Ecosyst.* **2008**, *18*, 780–798. [CrossRef]
4. Wohl, E.; Bledsoe, B.P.; Jacobson, R.B.; Poff, N.L.; Rathburn, S.L.; Walters, D.M.; Wilcox, A.C. The natural sediment regime in rivers: Broadening the foundation for ecosystem management. *Bioscience* **2015**, *65*, 358–371. [CrossRef]
5. Gallen, S.F.; Pazzaglia, F.J.; Wegmann, K.W.; Pederson, J.L.; Gardner, T.W. The dynamic reference frame of rivers and apparent transience in incision rates. *Geology* **2015**, *43*, 623–626. [CrossRef]
6. Davis, W.M.; Harvard, C. The topographic maps of the United States. *Science* **1893**, *21*, 225–226. [CrossRef] [PubMed]

7. Winslow, A.; Office of the, G.S. The Osage River and its meanders. *Science* **1893**, 31–32. [CrossRef]
8. Whipple, K.X.; DiBiase, R.A.; Crosby, B.T. *Bedrock Rivers*; Elsevier Ltd.: Amsterdam, The Netherlands, 2013; Volume 9, ISBN 9780080885223.
9. Sklar, L.S.; Dietrich, W.E. A mechanistic model for river incision into bedrock by saltating bed load. *Water Resour. Res.* **2004**, *40*. [CrossRef]
10. Zhang, L.; Parker, G.; Stark, C.P.; Inoue, T.; Viparelli, E.; Fu, X.; Izumi, N. Macro-Roughness model of bedrock-alluvial river morphodynamics. *Earth Surf. Dyn.* **2015**, *3*, 113–138. [CrossRef]
11. Thomson, J.; Thomson, W. On the origin of windings of rivers in alluvial plains, with remarks on the flow of water round bends in pipes. *Proc. R. Soc. Lond.* **1877**, *25*, 5–8. [CrossRef]
12. Fernández, R.; Parker, G.; Stark, C.P. Experiments on patterns of alluvial cover and bedrock erosion in a meandering channel. *Earth Surf. Dyn.* **2019**, *7*, 949–968. [CrossRef]
13. Ippen, A.T.; Drinker, P.A. Boundary Shear Stresses in Curved Trapezoidal Channels. *J. Hydraul. Div.* **1962**, *88*, 143–180. [CrossRef]
14. Yen, B.-C. *Characteristics of Subcritical Flow in a Meandering Channel*; The University of Iowa: Iowa City, IA, USA, 1965.
15. Graham, D.I.; James, P.W.; Jones, T.E.R.; Davies, J.M.; Delo, E.A. Measurement and Prediction of Surface Shear Stress in Annular Flume. *J. Hydraul. Eng.* **1992**, *118*, 1270–1286. [CrossRef]
16. Krishnappan, B.G. Rotating circular flume. *J. Hydraul. Eng.* **1993**, *119*, 758–767. [CrossRef]
17. Krishnappan, B.G.; Stone, M.; Granger, S.J.; Upadhayay, H.R.; Tang, Q.; Zhang, Y.; Collins, A.L. Experimental investigation of erosion characteristics of fine-grained cohesive sediments. *Water* **2020**, *12*, 1511. [CrossRef]
18. Sheng, Y.P. Consideration of Flow in Rotating Annuil for Sediment Erosion and Deposition Studies. *J. Coast. Res.* **1989**, *SPI 5*, 207–216.
19. Yang, Z.; Baptista, A.; Darland, J. Numerical modeling of flow characteristics in a rotating annular flume. *Dyn. Atmos. Ocean.* **2000**, *31*, 271–294. [CrossRef]
20. Desaulniers, G.; Frenette, M. Contributions to the Study of Erosion and Deposition Mechanisms in River Bends. In *Sedimentation, Symposium to Honor Prof. H. A. Einstein*; Hsieh Wen Shen 1972; pp. 10.1–10.19. Available online: https://www.oieau.org/eaudoc/system/files/documents/39/196460/196460_doc.pdf (accessed on 12 June 2021).
21. Engelund, F. Instability of flow in a curved alluvial channel. *J. Fluid Mech.* **1975**, *72*, 145–160. [CrossRef]
22. Ikeda, S. On secondary flow and dynamic equilibrium of transverse bed profile in alluvial curved open channel. *Proc. JSCE* **1974**, *1974*, 55–65. [CrossRef]
23. Kikkawa, H.; Ikeda, S.; Akira, K. Variation of bed profile with time in curved channel. *Proc. Jpn. Soc. Civ. Eng.* **1976**, 65–75. [CrossRef]
24. Baar, A.W.; de Smit, J.; Uijttewaal, W.S.J.; Kleinhans, M.G. Sediment Transport of Fine Sand to Fine Gravel on Transverse Bed Slopes in Rotating Annular Flume Experiments. *Water Resour. Res.* **2018**, *54*, 19–45. [CrossRef]
25. Booij, R. *Measurments of the Flow Field in a Rotating Annular Flume, Report No. 94-2*; TU Delft: Delft, The Netherlands, 1994. Available online: https://repository.tudelft.nl/islandora/object/uuid%3A431193bc-8cfb-46ce-81fd-5034941b0769 (accessed on 12 June 2021).
26. Fukuda, M.K.; Lick, W. The Entrainment of Cohesive Sediments in Freshwater. *J. Geophys. Res.* **1980**, *85*, 2813–2824. [CrossRef]
27. Taguchi, S.; Ozawa, H.; De Lima, A.C.; Izumi, N. Experimental study on bedrock incision and bed configuration in annular flume flow. *J. JSCE Hydraul. Eng.* **2017**, *73*, I_847–I_852. [CrossRef]
28. Taguchi, S.; Kobayashi, S.; De Lima, A.C.; Izumi, N. Experimental study on mixed bedrock-alluvial bed erosion in annular flume flow. *J. JSCE Hydraul. Eng.* **2018**, *74*, I_1039–I_1044. [CrossRef]
29. NORITAKE Co., Ltd. Noritake Products—Catalog Lineup. Available online: https://www.noritake.co.jp/eng/products/support/detail/11/ (accessed on 3 January 2022).
30. Johnson, J.P.; Whipple, K.X. Feedbacks between erosion and sediment transport in experimental bedrock channels. *Earth Surf. Process. Landf.* **2007**, *32*, 1048–1062. [CrossRef]
31. Shepherd, R.G. Incised River Meanders: Evolution in Simulated Bedrock. *Am. Assoc. Adv. Sci.* **1972**, *178*, 409–411. [CrossRef]
32. Karni, J.; Karni, E. Gypsum in construction: Origin and properties. *Mater. Struct.* **1995**, *28*, 92–100. [CrossRef]
33. Tohoku Keisya Co., Ltd. Tohoku Silica Sand Co. Available online: http://www.tohoku-keisya.co.jp/ (accessed on 3 January 2022).
34. Keyence Corporation CMOS Multi-Function Analog Laser Sensor IL Series—Instruction Manual. Available online: https://www.keyence.com/products/measure/laser-1d/il/downloads/?mode=ma&modelId=PM_235300 (accessed on 21 June 2021).
35. 3D Artec Artec 3D EVA. Available online: https://www.artec3d.com/portable-3d-scanners/artec-eva-v2 (accessed on 12 June 2021).
36. Allen, D.W. *GIS Tutorial 2: Spatial Analysis Workbook*; Esri Press: Redlands, CA, USA, 2016; ISBN 9781589484535.
37. Demeter, G.; Sklar, L.; Davis, J. The Influence of Variable Sediment Supply and Bed Roughness on the Spatial Distribution of Incision in a Laboratory Bedrock Channel. *AGU Fall Meet. Abstr.* **2005**, *86*. Available online: https://www.researchgate.net/publication/253214892_The_influence_of_variable_sediment_supply_and_bed_roughness_on_the_spatial_distribution_of_incision_in_a_laboratory_bedrock_channel (accessed on 12 June 2021).

water

Article

Extending the Applicability of the Meyer–Peter and Müller Bed Load Transport Formula

Epaminondas Sidiropoulos [1,*], Konstantinos Vantas [1], Vlassios Hrissanthou [2] and Thomas Papalaskaris [2]

[1] Faculty of Engineering, Aristotle University of Thessaloniki, 54124 Thessaloniki, Greece; kon.vantas@gmail.com
[2] Department of Civil Engineering, Democritus University of Thrace, 67100 Xanthi, Greece; vhrissan@civil.duth.gr (V.H.); tpapalas@civil.duth.gr (T.P.)
* Correspondence: nontas@topo.auth.gr

Abstract: The present paper deals with the applicability of the Meyer–Peter and Müller (MPM) bed load transport formula. The performance of the formula is examined on data collected in a particular location of Nestos River in Thrace, Greece, in comparison to a proposed Enhanced MPM (EMPM) formula and to two typical machine learning methods, namely Random Forests (RF) and Gaussian Processes Regression (GPR). The EMPM contains new adjustment parameters allowing calibration. The EMPM clearly outperforms MPM and, also, it turns out to be quite competitive in comparison to the machine learning schemes. Calibrations are repeated with suitably smoothed measurement data and, in this case, EMPM outperforms MPM, RF and GPR. Data smoothing for the present problem is discussed in view of a special nearest neighbor smoothing process, which is introduced in combination with nonlinear regression.

Keywords: bed load transport; random forests; Gaussian processes regression; Meyer–Peter and Müller formula; sediment transport

Citation: Sidiropoulos, E.; Vantas, K.; Hrissanthou, V.; Papalaskaris, T. Extending the Applicability of the Meyer–Peter and Müller Bed Load Transport Formula. *Water* **2021**, *13*, 2817. https://doi.org/10.3390/w13202817

Academic Editor: Wencheng Guo

Received: 30 August 2021
Accepted: 5 October 2021
Published: 11 October 2021

1. Introduction

Meyer–Peter, Favre and Einstein [1] published a formula in 1934 related to the transport of uniform sediment on a plane bed, while Meyer–Peter and Müller [2,3] published in 1948 and 1949 the definitive formula related to the transport of sediment mixtures with different values of specific gravity. The main characteristic of the Meyer–Peter and Müller (MPM) formula is the distinction of bed roughness due to individual particles from that bed roughness due to bed forms or the distinction of bed resistance due to skin friction from bed resistance due to bed forms. The historical development of the MPM formula is described in detail in Hager and Boes [4]. In this formula, the unit submerged sediment discharge is calculated, and the roughness effect of the channel bottom and walls is taken into account. Wong and Parker [5], by using the same databases of Meyer–Peter and Müller, have suggested two substantially revised forms of the MPM (1948) formula, in which no correction for bed forms is made. According to Wong and Parker [5], the form drag correction of the MPM formula is unnecessary in the context of the plane bed transport data. The amended bed load transport relations of Wong and Parker are valid for lower-regime plane bed equilibrium transport of uniform sediment.

In the formula applied in our study, the unit sediment discharge is calculated, while the channel bottom roughness is distinguished into roughness due to individual particles and roughness due to bed forms.

Herbertson [6] has examined the bed load formula of Meyer–Peter and Müller (1948) as well as other conventional bed load formulas using similitude theory as a common basis of comparison. Especially for wide channels with invariable grain size and ratio of sediment density to water density, Herbertson [6] suggests that the MPM formula is still incomplete in that the depth effect is not included. The final conclusion of Herbertson regarding the

MPM formula is cited as: " … the Meyer–Peter and Müller (1948) formula applies only to material rolling or sliding along the bed load and not to material in suspension, however temporarily. The latter condition would limit the formula to the lower regime of transport and presumably the formula will not take account of material transported by saltation".

Gomez and Church [7] have tested twelve bed load sediment transport formulas for gravel bed channels, among which is the MPM (1948) formula, using four sets of river data and three sets of flume data. On the basis of the tests performed, which were conducted in each case as if no sediment transport information were available for the river, none of the selected formulas and no other formula is capable of generally predicting bed load transport in gravel bed rivers.

Reid et al. [8] assessed the performance of several popular bed load formulas in the Negev Desert, Israel, and found that the Meyer–Peter and Müller [2] and Parker [9] equations performed best, but their analysis considered only one gravel bed river (Barry et al. [10]).

Martin [11] took advantage of ten years of sediment transport and morphologic surveys on the Vedder River, British Columbia, to test the performance of the Meyer–Peter and Müller [2] equation and two variants of the Bagnold equation [12]. The author concluded that the formulas generally under-predicted gravel transport rates [10].

The MPM formula [2] was also tested in comparison with the formulas of Parker [13], Schoklitsch [14] and Recking [15], by means of a field data set of 6319 bed load samples from sand and gravel bed rivers in the USA. The Meyer–Peter and Müller as well as the Parker equations were chosen because they permit a surface-based calculation with limited knowledge of sediment characteristics, and they are widely used [15]. The discrepancy ratio (average percentage of predicted bed load discharge not exceeding a factor of two in relation to the observed bed load discharge) obtained the value of 3% for the MPM formula, which is the lowest value in comparison to the corresponding values of the other three formulas.

López et al. [16] have tested the predictive power of ten bed load formulas against bed load rates for a large, regulated gravel bed river (Ebro River, NE Iberian Peninsula). The bed load MPM formula [2,3] was included in the ten bed load formulas tested. The discrepancy ratio, as it was defined above, was one of the formula's performance criteria applied. Especially for the MPM formula, the discrepancy ratio obtained the value 3% for the case of using surface bed material.

Overall, the predictive power of the MPM formula was relatively low. The MPM formula [2] belongs to that category of bed load formulas which are based on the assumption of a critical situation characterizing the incipient motion of grains on the bed. According to Meyer–Peter and Müller, the dimensionless critical shear stress amounts to 0.047. The same critical size for rough, turbulent flow, according to Shields [17], amounts to about 0.06. Gessler [18] reported a value of about 0.046 for a 50% probability of grain movement in a rough, turbulent flow. Miller et al. [19] arrived at a similar value of about 0.045 for rough, turbulent flow without consideration of the probability of movement.

Yang [20] suggested a dimensionless unit stream power equation for the computation and prediction of total sediment concentration without using any criterion for incipient motion. This equation was compared with a similar dimensionless unit stream power equation proposed by Yang [21], with the inclusion of criteria for incipient motion. In accordance with the comparison results, both equations are equally accurate in predicting the total sediment concentration in the sand size range. It should be noted that the new equation of Yang [20] is valid for sediment concentration greater than 20 ppm by weight. Both of Yang's equations were calibrated especially for Nestos River (northeastern Greece) on the basis of available measurements for bed load and suspended load. Regarding the comparison between predicted and measured values of total load transport rate, the values of the statistical criteria used for both equations were very satisfactory, as reported by Avgeris et al. [22].

Several studies have shown that omitting the incipient motion criterion may lead to better results, compared to the existing formulas. For example, Barry et al. [10], on the basis of 2104 bed load transport observations in 24 gravel bed rivers in Idaho (USA), concluded that formulas containing a transport threshold typically exhibit poor performance. Kitsikoudis et al. [23] have employed data-driven techniques, namely artificial neural networks, adaptive-network-based fuzzy inference system and symbolic regression based on genetic programming, for the prediction of bed load transport rates in gravel-bed steep mountainous streams and rivers in Idaho (USA). The derived models generated results superior to those of some of the widely used bed load formulas, without the need to set a threshold for the initiation of motion, and consequently avoid predicting erroneous zero transport rates.

Some previous studies of the authors on the calibration of MPM formula are reported below:

Papalaskaris et al. [24] have attempted to calibrate the MPM formula both manually and on the basis of the least squares method, in terms of roughness coefficient, for two streams in northeastern Greece: Kosynthos River and Kimmeria Torrent. Papalaskaris et al. [25] have also manually calibrated the MPM formula, in terms of roughness coefficient, for Nestos River (northeastern Greece). In a following study, Sidiropoulos et al. [26] have calibrated the same formula for Nestos River by means of a nonlinear optimization of two suitable parameters, while utilizing the average value of the roughness coefficient found by the manual calibration. In all three studies, the comparison between calculations and measurements of bed load transport rate was made on the basis of the following statistical criteria: root mean square error, relative error, efficiency coefficient, linear correlation coefficient, determination coefficient and discrepancy ratio. The values of the above statistical criteria for the case of manual calibration were more satisfactory, compared to the case of nonlinear optimization. However, the manual calibration was carried out on partial measurement sets, while the nonlinear optimization was carried out on the whole measurement set.

In view of the fact that the predictive power of the MPM formula did not reach particularly high levels, the present paper proposes an Enhanced MPM (EMPM) formula, demonstrating that, under suitable calibration, it shows a much better fitness to field data. Moreover, the performance of the enhanced formula is compared to machine learning methods, showing the competitiveness of the semi-empirical formula versus purely data-driven approaches.

One of the adjustment parameters of the formula is the critical shear stress, the value of which has been discussed and re-adjusted by various researchers, as already cited. In line with these investigations and under the data of this study, a zero value of this parameter gave the optimal calibration results.

Calibration of the EMPM formula was also performed on smoothed data, with the prospect of mitigating possible noise of the field measurements. A nearest neighbor smoothing is introduced and applied both in regard to the MPM formula calibration and to typical machine learning methods. In the case of the smoothed data, the performance of the EMPM formula turns out to be superior.

The introduction of machine learning methods into the field of sediment transport modeling brought about new standards in the error metrics of predicted versus measured data, sometimes tending to overshadow physically based and semi-empirical equations. This paper aims at turning attention back to such equations by introducing generalized forms and by establishing competitive and, in the case of smoothing, even better performance versus machine learning methods.

2. Materials and Methods

2.1. Study Area and Data

The Nestos River basin (Figure 1, northeastern Greece) considered in this study drains an area of 838 km^2 and lies downstream of Platanovrysi Dam. The river basin outlet is located at Toxotes. The river basin terrain is covered by forest (48%), bush (20%), cultivated land (24%), urban area (2%) and no significant vegetation (6%). The altitude ranges between 80 m and 1600 m, whereas the length of Nestos River is 55 km. The mean slope of Nestos River in the basin is 0.35%. The stream flow rate and bed load transport rate measurements concerning Nestos River were conducted at a location between the outlet of Nestos River basin (Toxotes) and the river delta. The measurement procedures are described in [25].

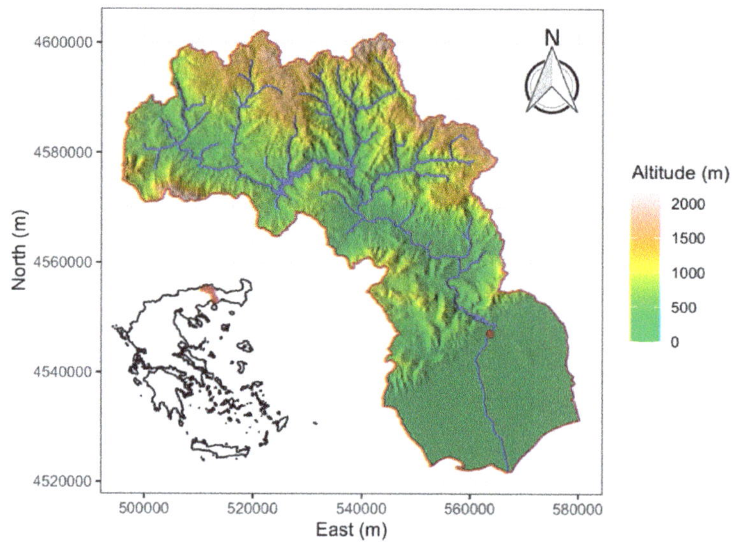

Figure 1. The Nestos River basin. The red filled circle symbolizes the location of the bed load measurements.

The first four statistical moments (mean, standard deviation, skewness and kurtosis) and other statistical properties were used to describe the bed load measured related values (Table 1).

Table 1. The average statistical properties of bed load related values. SD is an abbreviation for standard deviation.

Variable	Min	Mean	Median	Max	SD	Skew	Kurtosis
m_{Gm} (kg/(s·m))	0	0.0225	0.0175	0.0883	0.0201	1.0790	0.7741
Q (m^3/s)	0.100	2.457	1.760	11.020	2.242	1.872	3.393
b (m)	6.00	16.20	15.70	32.00	7.35	0.40	−0.82
h (m)	0.10	0.32	0.31	0.61	0.11	0.33	−0.21
u_m (m/s)	0.20	0.47	0.44	1.48	0.21	1.92	5.02
d_{90} (m)	0.0014	0.0028	0.0030	0.0037	0.0006	−0.8150	−0.2193
d_{50} (m)	0.0008	0.0045	0.0014	0.0235	0.0064	1.6859	1.0979

In concrete terms, m_{Gm} (kg/(s m)) is the measured bed load transport rate per unit width, Q (m^3/s) is the measured stream discharge, b (m) is the measured width of the assumed rectangular cross section, h (m) is the measured flow depth, u_m (m/s) is the measured mean flow velocity, d_{50} (m) is the median grain diameter of bed load, determined by the granulometric curves, and d_{90} (m) is a characteristic grain size diameter (in case of taking a stream bed load sample, concerning the sample weight, 90% is composed of grains with size less than or equal to d_{90}).

2.2. Meyer–Peter and Müller (MPM) Bed Load Transport Formula

In the MPM formula, referred to in the introduction, the unit submerged sediment discharge is calculated and the roughness effect of the channel bottom and walls is taken into account.

In the formula, as applied in our study, the unit sediment discharge is calculated while the channel bottom roughness is distinguished into roughness due to individual particles and roughness due to bed forms:

$$m_{Gc} = \frac{8}{g}\frac{\rho_F}{\rho_F - \rho_W}\sqrt{\frac{1}{\rho_W}}(\tau_o - \tau_{o,cr})^{3/2} \quad (1)$$

where:

$$\tau_o = \rho_W g I_r R_s,\ \tau_{o,cr} = 0.047\rho' \rho_W g d_m,\ \rho' = \frac{\rho_F - \rho_W}{\rho_W},\ I_r = \left(\frac{k_{st}}{k_r}\right)^{3/2} I \text{ and } k_r = \frac{26}{\sqrt[6]{d_{90}}} \quad (2)$$

The symbols of Equations (1) and (2) are explained below:

m_{Gc}: computed bed load transport rate per unit width (kg/(s·m))

g: gravity acceleration (m/s^2)

ρ_F: sediment density (kg/m^3)

ρ_w: water density (kg/m^3)

τ_o: actual shear stress (N/m^2)

$\tau_{o,cr}$: critical shear stress (N/m^2), characterizing the incipient motion of bed grains

I_r: energy line slope due to individual grains

R_s: hydraulic radius of the specific part of the cross section under consideration which affects the bed load transport (m).

d_m: mean diameter of bed load grains (m)

k_{st}: Strickler coefficient, the value of which depends on the roughness due to individual grains, as well as to stream bed forms ($m^{1/3}/s$).

k_r: coefficient, with value depending on the roughness due to individual grain ($m^{1/3}/s$)

I: energy line slope due to individual grains and stream bed forms

d_{90}: characteristic grain size diameter (m). It was defined for Table 1.

The basic limitations for the MPM formula are the following:

- Slope of energy line (I) from 0.04% to 2%
- Sediment particle size (d_{50}) from 0.4 mm to 20 mm
- Flow depth (h) from 0.01 m to 1.20 m
- Specific stream discharge (Q/b) from 0.002 m^2/s to 2 m^2/s
- Relative sediment density (ρ_F/ρ_w) from 0.25 to 3.2
- Particle size > 1 mm, to avoid the effects of apparent cohesion
- Flow depth > 0.05 m, to assure Froude similitude.

At this point, it should be noted that the mean values of the measured energy line slope (longitudinal bed slope in the case of uniform flow), sediment particle size, flow depth, specific stream discharge and relative sediment density are included in the ranges given above.

According to the Einstein–Barbarossa method (e.g., [27])

$$A = A_s + A_w = R_s U_s + R_w U_w \quad (3)$$

where A (m^2) is the stream cross section, assuming a rectangular section, approximately, and where R (m) is the hydraulic radius and U (m) is the wetted perimeter. The indices s and w stand for bed and walls, respectively. The hydraulic radius R_w is given by the familiar Manning formula:

$$R_w = \left(\frac{u_m}{k_w I^{0.5}}\right)^{1.5} \quad (4)$$

where u_m (m/s) is the mean flow velocity through the cross-sectional area A and k_w ($m^{1/3}$/s), a coefficient depending on the roughness of the walls. It is assumed that $k_w = k_{st}$. Additionally, I is set equal to the longitudinal stream bed slope on the basis of the assumption of uniform flow.

Then R_s, by combining Equations (3) and (4), turns out as

$$R_s = \frac{A - \left(\frac{u_m}{k_w I^{0.5}}\right)^{1.5} U_w}{U_s} \tag{5}$$

Equation (1) can be converted to a non-dimensional form (see Appendix A):

$$m^*_{Gc} = 8(1+\rho')\left[\left(\frac{k_{st}}{k_r}\right)^{3/2}\frac{Re^{*2}}{Re^2_{p50}} - 0.047\frac{d_m}{d_{50}}\right]^{3/2} \tag{6}$$

where

$$m^*_{Gc} = \frac{m_{Gc}}{\rho_w\sqrt{\rho' g d_{50}}\, d_{50}} \tag{7}$$

$$Re_{p50} = \frac{\sqrt{\rho' g d_{50}}\, d_{50}}{\nu} \tag{8}$$

$$Re^* = \frac{\sqrt{g R_s I}\, d_{50}}{\nu} \tag{9}$$

where ν is the kinematic viscosity of water. The size m_{Gc} becomes dimensionless by means of Equation (7). The derivation of Equation (9) is given in Appendix A.

Due to the sandy composition of the bed load in the river locations, the mean grain diameter d_m can be approximated by the median grain diameter d_{50}. Therefore, Equation (6) acquires the simpler non-dimensional form:

$$m^*_{Gc} = 8(1+\rho')\left[\left(\frac{k_{st}}{k_r}\right)^{3/2}\frac{Re^{*2}}{Re^2_{p50}} - 0.047\right]^{3/2} \tag{10}$$

The above non-dimensional scheme is in accordance with the dimensional analysis of Parker and Anderson [28], as utilized in a related paper by Kitsikoudis et al. [23]. Indeed, the non-dimensional groups appearing in Equation (10) are consistent with the dimensionless variables envisaged in the Parker and Anderson analysis. An analogous non-dimensional form has been presented by Wong and Parker [5], attributed originally to N. Chien in a 1954 publication of the US Army Corps of Engineers.

From the non-dimensional form (10), it turns out that, in a way compatible with the dimensional analysis of [28], the following non-dimensional variables determine bed load transport: Re_{p50} (Equation (8)), an explicit particle Reynolds number, Re^* (Equation (9)), a shear Reynolds number, ρ' (appearing in the third one of Equations (2)), the submerged specific gravity of the sediment.

2.3. Calibration of an Enhanced Meyer–Peter and Müller (EMPM) Formula

The available data comprise measured values for the physical parameters A, u_m, U_w, U_s and d_{50}, as well as measured values of bed transport rates m_{Gm}, denoted respectively as A_i, u_{mi}, U_{wi}, U_{si}, $d_{50,i}$ and m_{Gmi}, for i = 1, 2, . . . , N, where N is the number of data points. These subscripted quantities are substituted into the corresponding Equations (7)–(10), giving the non-dimensional bed load transport rate of Equation (10) in terms of measured quantities:

$$m^*_{Gci}(k_{st}) = 8(1+\rho')\left[\left(\frac{k_{st}}{k_{ri}}\right)^{3/2}\frac{Re_i^{*2}}{Re^2_{p50,i}} - 0.047\right]^{3/2} \tag{11}$$

In Equation (11), k_{st} emerges as an adjustment parameter. Therefore, the following expression can be used for the calibration of the MPM formula:

$$f(k_{st}) = \sum_{i=1}^{N} \left(m^*_{Gci}(k_{st}) - m^*_{Gmi}\right)^2 \quad (12)$$

where

$$m^*_{Gmi} = \frac{m_{Gmi}}{\rho_w \sqrt{\rho' g d_{50,i}}\, d_{50,i}},\ i = 1, \ldots, N \quad (13)$$

and m_{Gmi}, i = 1, … ,N, denote measured values of bed load transport rate. Calibration with respect to one parameter only, namely k_{st}, has already been tried (Sidiropoulos et al., 2018). In this paper, Equation (12) is further extended, so as to include more adjustment parameters:

$$m^*_{Gci}(k_{st},\ \alpha, \beta, \gamma) = 8\alpha(1 + \rho')\left[\left(\frac{k_{st}}{k_{ri}}\right)^{\beta} \frac{Re_i^{*2}}{Re^2_{p50,i}} - \gamma\right]^{3/2} \quad (14)$$

Equation (14) will be referred to as the Enhanced Meyer–Peter and Müller (EMPM) formula.

Equation (14) can be written as follows in a generalized form:

$$m^*_{Gci} = f_M(\mathbf{d}_M; \mathbf{p}_i) \quad (15)$$

where $\mathbf{d}_M = (k_{st}, \alpha, \beta, \gamma)$ is the vector of parameters,

$$\mathbf{p}_i = (A_i,\ U_{mi}, U_{wi}, U_{si}, d_{50,i}, k_{ri}, I_i),\ i = 1, 2, \ldots, N \quad (16)$$

is the vector of measured quantities that were suitably grouped in Equation (10), and

$$f_M(\mathbf{d}_M; \mathbf{p}_i) = 8\alpha(1 + \rho')\left[\left(\frac{k_{st}}{k_{ri}}\right)^{\beta} \frac{Re_i^{*2}}{Re^2_{p50,i}} - \gamma\right]^{3/2} \quad (17)$$

In analogy to Equations (11) and (12), the difference between computed and measured quantities is defined as

$$e_{iM} = f_M(\mathbf{d}_M; \mathbf{p}_i) - m^*_{Gmi} \quad (18)$$

where m^*_{Gmi} is given by Equation (13) and f_M by Equation (17).

The objective function of the calibration problem is

$$F_M(\mathbf{d}_M) = \sum_{i=1}^{N} e^2_{iM} \quad (19)$$

where N is the number of measurement points.

The minimization of the objective function F_M of Equation (19) was executed by a genetic algorithm followed by a Nelder–Mead local search.

2.4. Application of Machine Learning Schemes

A common pitfall in the use of machine learning algorithms is the expectation of performance regardless of the nature and limitations of the problem and of the presence of noisy data. In the context of bed load estimation, especially for data that are coming from natural streams, errors are expected to be higher, as testified by the sediment transport literature, in which errors are predominantly reported as ratios and not as differences of compared quantities.

In a preliminary stage of analysis, various machine and statistical learning methods have been evaluated, such as neural networks (various architectures and regularization techniques), support vector regression, decision trees, linear models, K-NN regression,

Gaussian Processes Regression (GPR) and Random Forests (RF). Most methods gave similar results with the exception of neural networks, which had a tendency to overfit or, in other words, they fitted too closely on the data, memorizing the noise and, as a result, were unable to adequately generalize on new data.

In the sequel, RF and GPR algorithms are presented so as to have a broader representation of machine learning methods. RF had the best performance and the results obtained regarding their generalization ability in this problem and dataset justify their use, as reported later in Section 3.

2.4.1. Random Forests

Random Forests (RF) is a data-driven algorithm in the area of supervised learning which tries to fit a model using a set of paired input variables and their associated output responses, and can be used in classification and regression problems. In summary, RF consists of a number of decision trees [29]. For each tree, a random set is created from the dataset via bootstrapping [29], and in each node of the tree a random set of n input variables from the p variables of the dataset is considered to pick the best split [29]. The prediction of the output response in regression problems is the mean value of the estimations of these random decision trees. RF is one of the most popular methods applied in machine learning because of: (a) its robustness to outliers and overfitting, (b) its ability to perform feature selection and (c) the fact that its default hyperparameters (i.e., the set of parameters that have to be selected a priori in order to train a RF), as implemented in software, give satisfactory results [30].

The measured data quantities defined above, under the vector $\mathbf{p}_i$, serve as input variables in the RF learning scheme, while m_{Gmi} will be the corresponding target for the output. The general form of Equations (18) and (19) can be used again for the formation of the objective function, as follows:

Let

$$m^*_{Gci} = f_R(\mathbf{d}_R; \mathbf{p}_i) \quad (20)$$

in analogy to Equation (15), where $\mathbf{d}_R$ is the vector of the RF parameters (i.e., the set of decision trees that operate as an ensemble) that will be determined through training.

Then the deviation of measured from computed values is

$$e_{iR} = f_R(\mathbf{d}_R; \mathbf{p}_i) - m^*_{Gmi} \quad (21)$$

and the objective value of the problem will be

$$F_R(\mathbf{d}_R) = \sum_{i=1}^{N} e^2_{iR} \quad (22)$$

2.4.2. Gaussian Processes Regression

A completely different alternative is Gaussian Processes Regression (GPR), which can be briefly described as follows: In the area of machine learning, Williams and Rasmussen [31] developed a regression algorithm based on Gaussian processes, which is a non-parametric, Bayesian approach that has the ability to work well with small datasets. In the framework of that algorithm, the prediction for an input test point $\mathbf{x}$ is derived by means of a Gaussian stochastic process with an assumed mean equal to 0 and with a variance σ^2 calculated in terms of covariances involving $\mathbf{x}$ and the training data. A suitable covariance function is selected and parametrized, and finally, the hyperparameters involved are determined through optimization.

In this case, the formal scheme of Equations (20)–(22) is retained, with $\mathbf{d}_R$ replaced by $\mathbf{d}_G$ for GPR. The vector $\mathbf{d}_G$ represents the internal parameters of the respective machine learning process, which will be optimally determined according to the above outline.

2.5. Training and Testing Procedures

Three methods are presented here for modeling bed load sediment transport. The first one is the calibration of the EMPM formula, while the second and the third consist in the application of RF and GPR machine learning methods respectively. In all three cases, a resampling method needs to be executed in order to estimate the generalization error of the methods or, specifically, the measure of accuracy of the methods to predict outcome values from data that are not known a priori. For that reason, bootstrapping [29] is applied, a procedure that was repeated 100 times.

Every bootstrap sample dataset consists of 116 points generated through random sampling with replacement of the original 116 data points. Consequently, some observations may appear more than once and some not at all. The latter are used to estimate the generalization error (out-of-the-bag error) and the former lead to the training error of the methods.

The simulation will be carried out for the original raw data, first by calibration of the EMPM formula and then by training the RF and GPR machine learning methods. Thereafter, the original data will be subjected to smoothing and the simulation will be repeated for the smoothed data by the same three methods.

2.6. Nearest Neighbor Smoothing

A smoothing process based on nearest neighbors is introduced as follows:

For each vector $\mathbf{p}_i$ of input measured quantities (Equation (16)), the distances are computed to all other vectors $\mathbf{p}_j$, j = 1, 2, … , N, where N is the number of available measurement points. The k nearest neighbors to $\mathbf{p}_i$ are then picked out and the average of these is taken, as well as the average of the corresponding bed load transport measurements. These averages will replace the original data. The process is formalized as follows:

Let

$$r_{i,j} = ||\mathbf{p}_i - \mathbf{p}_j||,\ j = 1,\ 2, \ldots, N \tag{23}$$

be the distances between $\mathbf{p}_i$ and all $\mathbf{p}_j$'s, including $\mathbf{p}_i$ itself.

Let

$$D_i = \{r_{i1}, r_{i1}, \ldots, r_{iN}\} \tag{24}$$

be the set of the distances of p_i from all other parameter vectors, as computed according to Equation (22).

Let

$$r_{i,i1}, r_{i,i2}, \ldots . r_{i,ik}$$

be the k smallest members of the set D_i of Equation (23) and let

$$\mathbf{p}_{i1}, \mathbf{p}_{i2}, \ldots, \mathbf{p}_{ik} \text{ and } m^*_{Gmi1}, m^*_{Gmi2}, \ldots, m^*_{Gmik}$$

be the corresponding k vectors and associated bed load measurements expressed in non-dimensional form as above (Equation (13)).

Then the following averaging is performed:

$$\mathbf{p}_i^k = \frac{1}{k}\sum_{j=i}^{k} \mathbf{p}_{ij},\ i = 1, 2, \ldots, N \tag{25}$$

$$m^{*k}_{Gmi} = \frac{1}{k}\sum_{j=i}^{k} m^*_{Gmij},\ i = 1, 2, \ldots, N \tag{26}$$

The pair $(\mathbf{p}_i^k, m^{*k}_{Gmi})$ will replace the pair $(\mathbf{p}_i, m^*_{Gmi})$ in Equations (18) and (19) for the formation of objective function F_M, and in Equations (21) and (22) for the formation of objective function F_R.

It needs to be noted here that the nearest neighbor technique is used for smoothing only and not for prediction, as known from the literature (e.g., [29]). Prediction is performed by the nonlinear regression that follows the smoothing.

3. Results and Discussion

3.1. Implementation of Algorithms

The EMPM formula presented in this paper was calibrated with respect to the four parameters k_{st}, α, β and γ, contained in the objective function of Equation (19). The last parameter γ is related to the so-called threshold referred to in the introduction, which turned out to be zero for our data and for all repetitions of the EMPM formula calibrations. The required minimization was executed by a genetic algorithm followed by a Nelder–Mead local search. As is well-known, the genetic algorithm points to the area within which the minimum needs to be sought and the sequent local search more accurately yields the location of the minimum [32].

The RF and GPR applications are completely independent, pure data-driven processes, as already explained in Section 2.4. They were performed by means of the algorithms provided by the Mathematica Computation System [33], with automatic selection of the optimal hyperparameters.

For the nearest neighbor smoothing, the number of nearest neighbors was chosen equal to k = 5.

3.2. Error Metrics

The error metrics included are RMSE, Mean Discrepancy Ratio (MDR) and coefficient of determination (R^2).

The discrepancy ratio for individual measured–calculated quantities is defined as

$$dr_i = m_{Gci}/m_{Gmi}, \, i = 1, 2, \ldots, N \tag{27}$$

where N is the number of measurements.

The Mean Discrepancy Ratio (MDR) is

$$\mathrm{MDR} = \frac{1}{N}\sum_{i=1}^{N} dr_i \tag{28}$$

As already stated in Section 2.5, the process of training and test set formation was repeated 100 times within a bootstrapping scheme, each time obtaining a different set of the adjustment parameters α, β and γ and corresponding values of RMSE. Tables 2 and 3 give, for each case, the mean value of RMSE over the 100 repetitions, along with the associated standard deviation. The same tables also show the mean values and standard deviations for the 100 repetitions of RMSE, MDR and R^2.

Table 2. Mean and standard deviation of error metrics for raw data and for the out-of-the-bag dataset from bootstrap resampling. RMSE units: kg/(s·m).

		Mean	**SD**
RDGP Raw Data Original Formula	RMSE	**0.05646**	0.00564
	MDR	6.20336	1.44002
	R^2	0.16077	0.12808
RDEF Raw Data Enhanced Formula	RMSE	**0.01928**	0.00217
	MDR	3.06172	0.92239
	R^2	0.47922	0.08730

Table 2. *Cont.*

RDRF Raw Data Random Forests	RMSE	**0.01832**	0.00247
	MDR	3.49139	1.49874
	R^2	0.54513	0.10054
RDGP Raw Data Gaussian Processes	RMSE	**0.01999**	0.00280
	MDR	3.36799	1.82963
	R^2	0.34135	0.10937

Table 3. Mean and standard deviation of error metrics for smoothed data and for the out-of-the-bag dataset from bootstrap resampling. RMSE units: kg/(s·m).

		Mean	**SD**
SDOF	RMSE	**0.04159**	0.006065
Smoothed Data	MDR	2.37244	0.719034
Original Formula	R^2	−0.94236	0.235836
SDEF Smoothed Data Enhanced Formula	RMSE	**0.00926**	0.001925
	MDR	1.20260	0.27857
	R^2	0.48299	0.20773
SDRF Smoothed Data Random Forests	RMSE	**0.01030**	0.002019
	MDR	1.31873	0.344024
	R^2	0.40220	0.123037
SDGP Smoothed Data Gaussian Processes	RMSE	**0.01180**	0.002842
	MDR	1.27084	0.356768
	R^2	0.22199	0.172782

The comparisons among the employed algorithms will be based mainly on RMSE, which also formed the basis of the objective function. R^2 is a well-established index in hydrologic modeling, being the same as the well-known Nash–Sutcliff Efficiency index (NSE). However, serious reservations have been reported in the pertinent literature as to its autonomous use for performance evaluation [34]. Various cases have been reported in which a good model could have low R^2 and a bad model high R^2. Also, in the context of sediment transport modeling, due to inherent difficulties and due to the noise in field data, low R^2 values would not be a surprise.

In the same context and for the same reasons, the discrepancy ratio, rather than R^2, has been a primary realistic index, since it estimates errors in terms of ratios rather than differences between measured and computed values. Regarding discrepancy ratios in this study, a qualitative comparison of the present results to analogous results of the literature will be given in Section 3.4.

3.3. Results of Error Metrics and Related Comparisons

Table 2 covers the various cases based on raw data for EMPM vs. MPM and vs. RF and GPR, while Table 3 covers the corresponding cases for smoothed data. The nomenclature of the various cases is given on the leftmost columns of Tables 2 and 3.

As expected, the performance of the calibrated EMPM formula is definitely higher in comparison to the original MPM formula. Indeed, it can be seen in Table 2 that, for Raw Data and for the Enhanced MPM Formula (RDEF), the out-of-the-bag test set RMSE mean value for the 100 bootstrapping repetitions is equal to 0.019278, while the corresponding quantity for the Original Formula (RDOF) is equal to 0.056466.

It can be observed from Tables 2 and 3 that the standard deviations, especially for RMSE, are by an order of magnitude smaller than the mean values. Therefore, the mean error metrics of the tables and, especially, the mean RMSE, are representative.

Although it may not be of particular use at present, indicative sets of parameters can be given and, as such, those are chosen that closely produce the mean RMSE of RDEF and SDEF, respectively. Specifically,

- for RDEF, $k_{st} = 10.7379$, $\alpha = 0.2000$, $\beta = 1.1199$, $\gamma = 0$ and
- for SDEF, $k_{st} = 30.00$, $\alpha = 0.6519$, $\beta = 3.5374$, $\gamma = 0$.

While RDEF is superior to RDOF, it is also competitive versus the machine learning methods (RDRF and RDGP), as seen on Table 2. On the other hand, in the cases of smoothed data (Table 3), EMPM (SDEF) outperforms the original formula MPM (SDOF), as well as Random Forests (SDRF) and Gaussian Processes Regression (SDGP). These facts are seen clearly by observing the respective RMSE's in Tables 2 and 3, along with the corresponding small standard deviations.

As discussed in Section 3.2, the values of R^2 are not to be taken as a sole representative measure of performance. In comparative terms, the R^2 value for RDEF from Table 2 is 0.47962, while for RDOF it is 0.169077; i.e., R^2 is about three times greater for the enhanced versus the original MPM formula.

The same relative improvement in performance applies to RMSE, and the conclusions drawn from Table 2 can be visualized in Figure 2, which shows the corresponding boxplots.

Similar comparisons can be made for the other cases of Tables 2 and 3, but the general status of the results is further indicated by the comparison of discrepancy ratios obtained in this study versus analogous quantities of the pertinent literature, as described in the next section.

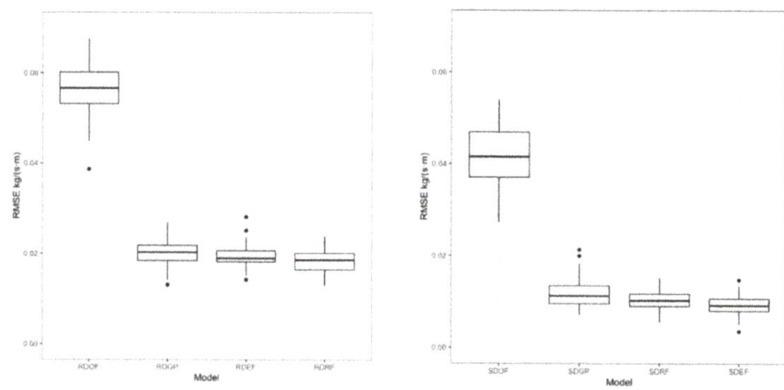

(a) (b)

Figure 2. Boxplots for: (**a**) out-of-the-bag RMSE for all methods and bootstrapping repetitions using raw data; (**b**) out-of-the-bag RMSE for all methods and bootstrapping repetitions using smoothed data.

3.4. Discrepancy Ratio Comparisons

In the context of sediment transport modeling, an established indicator is the discrepancy ratio. It would, therefore, be appropriate to consider characteristic results of this ratio in the pertinent literature and compare them to those of the present study. Indeed, in References [15,16], three indices, D_1, D_2 and D_3, appear in relation to discrepancy ratios:

Let dr_i's be the ratios defined in Equation (27). Then

1. D_1 is defined as the percentage of dr_i's such that $0.5 \leq dr_i \leq 2$.
2. D_2 is defined as the percentage of dr_i's such that $0.25 \leq dr_i \leq 4$.
3. D_3 is defined as the percentage of dr_i's such that $0.1 \leq dr_i \leq 10$.

In [15], a characteristic set (D_1, D_2, D_3) is equal to (3%, 7%, 9%), especially for the MPM formula resulting from 6319 values of a field dataset regarding sand and gravel bed streams in USA. In [16], D_1 = 3%, especially for the MPM formula applied to Ebro River (Spain) with the gravel bed.

Indicative computed corresponding values in this study are as follows:

1. Raw data, original formula: (D_1, D_2, D_3) = (32%, 45%, 60%).
2. Raw data, enhanced formula: (D_1, D_2, D_3) = (55%, 81%, 97%).

The above results are not directly comparable to those of [15,16] due to different datasets and validation schemes. However, they render a good indication of a better fitting and of the improvement brought about by the enhanced formula. Discrepancy ratio results in this study are even better for the smoothed data.

3.5. Examination of Possible Overfitting

In order to detect any cases of overfitting, Table 5 lists the RMSE mean values of the bootstrapped training sets versus those of the corresponding out-of-the-bag test sets. The latter are also given in Tables 2 and 3. It is immediately seen that there is no question of overfitting in the EMPM formula (RDEF and SDEF), while there is an indication of overfitting in the GPR (RDGP and SDGP) due to larger differences between the RMSE mean values of training and test sets. Regarding RF, there is clearly no overfitting in RDRF. In the case of the smoothed data (SDRF), overfitting to a lesser degree versus SDGP is indicated in Table 4.

Overfitting would not be expected of the EMPM formula, and this fact is verified in the above Table 4, but overfitting is very often not easily avoidable in machine learning methods. The best behavior in this regard was exhibited by RF.

Table 4. RMSE mean values of bootstrap training and out-of-the-bag test sets. RMSE units: kg/(s·m).

	Training Set	Test Set
RDOF	0.05427	0.05646
RDEF	0.01785	0.01928
RDRF	0.01367	0.01833
RDGP	0.00690	0.01999
SDOF	0.04476	0.04159
SDEF	0.01037	0.00926
SDRF	0.00696	0.01030
SDGP	0.00191	0.01180

3.6. Statistical Comparisons

Besides the computation results shown in Tables 2 and 3 and in Figure 2, statistical comparisons between the methods used are also in order.

The comparison of the algorithms was based on the work of Demšar [35] and García et al. [36] on the use of non-parametric methods for the evaluation of results of machine learning algorithms, because parametric hypothesis testing methods (pairwise *t*-test and ANOVA) were not deemed suitable due to the nature of the algorithms. The Friedman test [37] was performed in order to determine whether an algorithm has a systematically better or worse performance. The obtained *p*-values (both $< 2.2 \times 10^{-18}$) indicated that the null hypothesis of all the algorithms performing the same could safely be rejected. Then, post-hoc tests followed for all possible pairs of algorithms using the Wilcoxon signed rank test [38]. Because of the multiple pairwise tests, the *p*-values that resulted were adjusted using the Benjamini and Hochberg method [39], which controls the false discovery rate (Tables 5 and 6). Table 5 shows the adjusted *p*-values below the main diagonal and at the upper diagonal positions, showing the estimated differences between methods. Table 6 shows the corresponding quantities for smoothed data. Indeed, the *p*-values indicate statistical differences, and it is noted that the enhanced formula not only shows better performance compared to the original formula, but also compared to the machine learning methods.

Table 5. To the right of the diagonal stand the estimated differences of RMSE between models for raw data. To the left of the diagonal stand the adjusted *p*-values for the H_0 (null hypothesis): difference = 0.

	RDOF	RDGP	RDEF	RDRF
RDOF	-	0.0363	0.0377	0.0381
RDGP	7.9×10^{-18}	-	0.0014	0.0017
RDEF	7.9×10^{-18}	0.02	-	0.0003
RDRF	7.9×10^{-18}	4.3×10^{-5}	0.01	-

Table 6. To the right of the diagonal stand the estimated differences of RMSE between models for smoothed data. To the left of the diagonal stand the adjusted *p*-values for the H_0 (null hypothesis): difference = 0.

	SDOF	SDGP	SDRF	SDEF
SDOF	-	0.0303	0.0313	0.0323
SDGP	7.9×10^{-18}	-	0.0010	0.0020
SDRF	7.9×10^{-18}	8.2×10^{-4}	-	0.0010
SDEF	7.9×10^{-18}	7.6×10^{-9}	$5.3 \cdot 10^{-4}$	-

3.7. Indicative Scatter Plots

Figure 3 shows indicative scatter plots of predicted versus measured values for a random out-of-the-bag dataset for observed bed load measurement values (kg/(s·m)), and Figure 4 shows corresponding plots for smoothed data. It can visually be verified that the enhanced formula and both machine learning methods exhibit superior performance compared to the initial MPM formula, consistent with the mean error metrics of Tables 2 and 3.

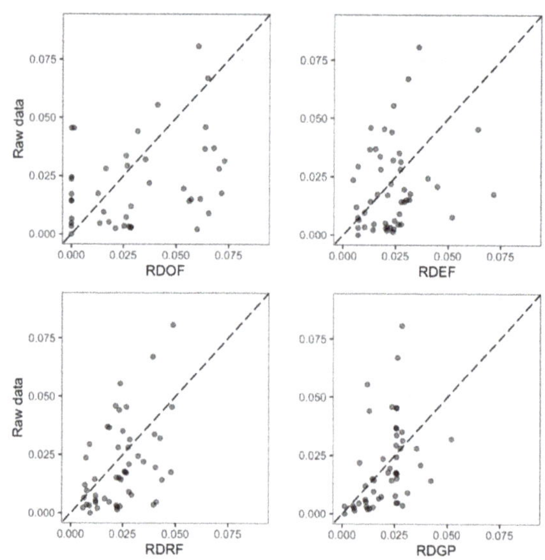

Figure 3. Predicted values for a random out-of-the-bag dataset using the optimal tuned parameters of each model versus observed values of bed load.

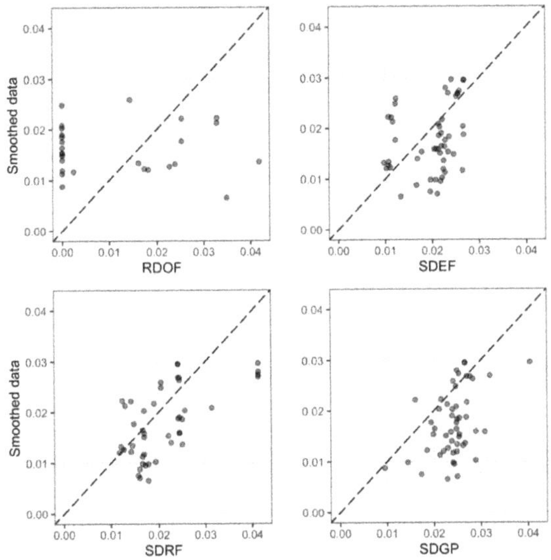

Figure 4. Predicted values for a random out-of-the-bag dataset using the optimal tuned parameters of each model versus smoothed values of bed load.

4. Conclusions

The advent of machine learning methods has also contributed to progress in the area of sediment transport modeling. Purely data-driven methods that appeared in the literature were found to outperform well-known physically based and semi-empirical equations. In an effort to enhance the performance of such equations, the Meyer–Peter and Müller bed load transport formula is extended in the present paper by the addition of suitable adjustment parameters, for the purpose of reinforcing its predictive abilities.

The resulting Enhanced Meyer–Peter and Müller formula presented a definitely improved performance in comparison to the original formula, one which is also competitive to purely data-driven techniques and even superior in the case of smoothed data. As a characteristic data-driven technique, the Random Forests learning scheme was chosen, due to its advantages in terms of robustness against outliers and overfitting, considering the noise contained in the field data of this study. A completely different machine learning method, Gaussian Processes Regression, was also tried and gave similar results, but was found to overfit on the training data.

For the purpose of countering noise effects, data smoothing is important and needs to be further considered for problems involving sediment transport field data, such as the present one. A nearest neighbor data smoothing process is presented and combined with nonlinear regression, a scheme different from the well-known nearest neighbor regression of the literature. Under smoothing, the enhanced MPM formula shows better performance, even compared to the machine learning methods.

The methods presented in this paper call for further applications in other natural streams with values of the variables beyond the range of boundary values given in the present article, as well as in modeling with laboratory data. Additionally, further research would be useful in the direction of hybrids involving both machine learning and sediment transport formulas.

Author Contributions: E.S. Conceptualization, algorithm development; K.V. Machine learning application, statistical evaluation, V.I.I. Physical problem formulation, general supervision, T.P. Data acquisition and preparation. All authors have read and agreed to the published version of the manuscript.

Funding: This research received no external funding.

Institutional Review Board Statement: Not applicable.

Informed Consent Statement: Not applicable.

Data Availability Statement: The data are listed in the Appendix A of Reference [26].

Conflicts of Interest: The authors declare no conflict of interest.

Appendix A

Appendix A.1. Derivation of Equation (6)

In Equation (1), the shear stresses τ_o and $\tau_{o,cr}$ are replaced by their values from the two first Equations (2):

$$m_{Gc} = \frac{8\rho_F}{g(\rho_F - \rho_w)\sqrt{\rho_w}}(\rho_w g I_r R_s - 0.047\rho'\rho_w g d_m)^{3/2} = \frac{8\rho_F \rho_w \sqrt{g}}{\rho_F - \rho_w}(I_r R_s - 0.047\rho' d_m)^{3/2} \quad \text{(A1)}$$

Both sides of Equation (A1) are divided by $\rho_w\sqrt{\rho' g d_{50}} d_{50}$ (d_{50}: median grain size):

$$\frac{m_{Gc}}{\rho_w\sqrt{\rho' g d_{50}} d_{50}} = \frac{8\frac{\rho_F}{\rho_w}}{\rho'^{3/2} d_{50}^{3/2}}(I_r R_s - 0.047\rho' d_m)^{3/2} \quad \text{(A2)}$$

or, by taking into account the fourth of Equations (2) and (7):

$$m^*_{Gc} = 8\frac{\rho_F}{\rho_w}[\left(\frac{k_{st}}{k_r}\right)^{\frac{3}{2}}\frac{IR_s}{\rho' d_{50}} - 0.047\frac{d_m}{d_{50}}]^{3/2} \quad \text{(A3)}$$

However, the following relationships are valid: $\frac{\rho_F}{\rho_w} = 1 + \rho'$ and, by taking into account Equations (8) and (9):

$$\frac{IR_s}{\rho' d_{50}} = \frac{Re^{*2}}{Re^2_{p50}} \quad \text{(A4)}$$

On the basis of Equations (A4) and (A3), it becomes:

$$m^*_{Gc} = 8(1+\rho')[\left(\frac{k_{st}}{k_r}\right)^{\frac{3}{2}}\frac{Re^{*2}}{Re^2_{p50}} - 0.047\frac{d_m}{d_{50}}]^{3/2} \quad \text{(A5)}$$

Appendix A.2. Derivation of Equation (9)

The shear Reynolds number (or sediment Reynolds number) Re^* is mathematically defined as follows:

$$Re^* = \frac{u * d_{50}}{\nu} \quad \text{(A6)}$$

$u*$: bed shear velocity (m/s)
d_{50}: median grain diameter (m)
ν: kinematic viscosity of water (m^2/s).
The definition of the shear Reynolds number is analogous to that of the classical Reynolds number Re:

$$Re = \frac{u_m h}{\nu} \quad \text{(A7)}$$

u_m: mean flow velocity (m/s)
h: flow depth (m).
The bed shear velocity u^* (m/s) is mathematically defined as follows:

$$u* = \sqrt{\frac{\tau_o}{\rho_w}} = \sqrt{\frac{\rho_w g R I}{\rho_w}} = \sqrt{gRI} \quad \text{(A8)}$$

τ_o: bed shear stress (N/m^2)
ρ_w: water density (kg/m^3)
g: gravitational acceleration (m/s^2)
R: hydraulic radius (m)
I: energy line slope.

In Equation (A8), the hydraulic radius R regarding the whole cross section is replaced by the hydraulic radius R_s regarding the specific part of the cross section which affects the bed load transport:

$$u* = \sqrt{gR_s I} \quad \text{(A9)}$$

So, Equation (A6) becomes:

$$Re^* = \frac{\sqrt{gR_s I}d_{50}}{\nu} \quad \text{(A10)}$$

References

1. Meyer-Peter, E.; Favre, H.; Einstein, A. Neuere Versuchsresultate über den Geschiebetrieb. *Schweiz. Bauztg.* **1934**, *103*, 147–150.
2. Meyer-Peter, E.; Müller, R. Formulas for bed-load transport. In Proceedings of the 2nd IAHR Congress, Stockholm, Sweden, 7–9 June 1948; IAHR: Delft, The Netherlands, 1948; Volume A2, pp. 1–26.
3. Meyer-Peter, E.; Müller, R. Eine Formel zur Berechnung des Geschiebetriebs. *Schweiz. Bauztg.* **1949**, *67*, 29–32.
4. Hager, W.H.; Boes, R.M. Eugen Meyer-Peter and the MPM sediment transport formula. *J. Hydraul. Eng.* **2018**, *144*, 02518001. [CrossRef]
5. Wong, M.; Parker, G. Reanalysis and correction of bed-load relation of Meyer-Peter and Müller using their own database. *J. Hydraul. Eng.* **2006**, *112*, 1159–1168. [CrossRef]
6. Herbertson, J.G. A critical review of conventional bed load formulae. *J. Hydrol.* **1969**, *8*, 1–26. [CrossRef]
7. Gomez, B.; Church, M. An assessment of bed load sediment transport formulae for gravel bed rivers. *Water Resour. Res.* **1989**, *25*, 1161–1186. [CrossRef]
8. Reid, I.; Powell, D.M.; Laronne, J.B. Prediction of bed-load transport by desert flash floods. *J. Hydraul. Eng.* **1996**, *122*, 170–173. [CrossRef]
9. Parker, G. Surface-based bed load transport relation for gravel rivers. *J. Hydraul. Res.* **1990**, *28*, 417–436. [CrossRef]
10. Barry, J.J.; Buffington, J.M.; King, J.G. A general power equation for predicting bed load transport rates in gravel bed rivers. *Water Resour. Res.* **2004**, *40*, W10401. [CrossRef]
11. Martin, Y. Evaluation of bed load transport formulae using field evidence from the Vedder River, British Columbia. *Geomorphology* **2003**, *53*, 73–95. [CrossRef]
12. Bagnold, R.A. An empirical correlation of bed load transport rates in flumes and natural rivers. *Proc. R. Soc. Lond.* **1980**, *372*, 453–473.
13. Parker, G. Hydraulic geometry of active gravel rivers. *J. Hydraul. Div.* **1979**, *105*, 1185–1201. [CrossRef]
14. Schoklitsch, A. *Handbuch des Wasserbaus*, 3rd ed.; Springer: Wien, Austria, 1962.
15. Recking, A. A comparison between flume and field bed load transport data and consequences for surface-based bed load transport prediction. *Water Resour. Res.* **2010**, *46*, W03518. [CrossRef]
16. López, R.; Vericat, D.; Batalla, R.J. Evaluation of bed load transport formulae in a large regulated gravel bed river: The lower Ebro (NE Iberian Peninsula). *J. Hydrol.* **2014**, *510*, 164–181. [CrossRef]
17. Shields, A. *Anwendung der Ähnlichkeitsmechanik und der Turbulenzforschung auf die Geschiebebewegung*; Mitteilungen der Preussischen Versuchsanstalt für Wasser-und Schiffbau, Heft 26; Berlin, Germany, 1936.
18. Gessler, J. Beginning and ceasing of sediment motion. In *River Mechanics*; Shen, H.W.: Fort Collins, CO, USA, 1971.
19. Miller, M.C.; McCave, I.N.; Komar, P.D. Threshold of sediment motion under unidirectional currents. *Sedimentology* **1977**, *24*, 507–527. [CrossRef]
20. Yang, C.T. Unit stream power equation for total load. *J. Hydrol.* **1979**, *40*, 123–138. [CrossRef]
21. Yang, C.T. Incipient motion and sediment transport. *J. Hydraul. Div.* **1973**, *99*, 1679–1704. [CrossRef]
22. Avgeris, L.; Kaffas, K.; Hrissanthou, V. Comparison between calculation and measurement of total sediment load: Application to Nestos River. *Environ. Sci. Proc.* **2020**, *2*, 19. [CrossRef]
23. Kitsikoudis, V.; Sidiropoulos, E.; Hrissanthou, V. Machine learning utilization for bed load transport in gravel-bed rivers. *Water Resour. Manag.* **2014**, *28*, 3727–3743. [CrossRef]
24. Papalaskaris, T.; Hrissanthou, V.; Sidiropoulos, E. Calibration of a bed load transport rate model in streams of NE Greece. *Eur. Water* **2016**, *55*, 125–139.
25. Papalaskaris, T.; Dimitriadou, P.; Hrissanthou, V. Comparison between computations and measurements of bed load transport rate in Nestos River, Greece. *Proc. Eng.* **2016**, *162*, 172–180. [CrossRef]

26. Sidiropoulos, E.; Papalaskaris, T.; Hrissanthou, V. Parameter Optimization of a Bed load Transport Formula for Nestos River, Greece. *Proceedings* **2018**, *2*, 627. [CrossRef]
27. Hrissanthou, V.; Tsakiris, G. Sediment Transport. In *Water Resources: I. Engineering Hydrology*; Tsakiris, G., Ed.; Symmetria: Athens, Greece, 1995; Chapter 16; pp. 537–577. (In Greek)
28. Parker, G.; Anderson, A.G. Basic principles of River Hydraulics. *J. Hydraul. Div.* **1977**, *103*, 1077–1087. [CrossRef]
29. James, G.; Witten, D.; Hastie, T.; Tibshirani, R. *An Introduction to Statistical Learning*; Springer: Berlin/Heidelberg, Germany, 2021.
30. Vantas, K.; Sidiropoulos, E.; Loukas, A. Estimating Current and Future Rainfall Erosivity in Greece Using Regional Climate Models and Spatial Quantile Regression Forests. *Water* **2020**, *12*, 687. [CrossRef]
31. Rasmussen, C.E.; Williams, K.I. *Gaussian Processes for Machine Learning*; MIT Press: Cambridge, MA, USA, 2006.
32. Chelouah, R.; Siarry, P. Genetic and Nelder-Mead algorithms hybridized for a more accurate global optimization of continuous multiminima functions. *Eur. J. Oper. Res.* **2003**, *148*, 335–348. [CrossRef]
33. *Mathematica*; Version 12.3.1; Wolfram Research, Inc.: Champaign, IL, USA, 2021.
34. Jain, S.K.; Sudheer, K.P. Fitting of Hydrologic Models: A Close Look at the Nash–Sutcliffe Index. *J. Hydrol. Eng.* **2008**, *13*, 981–986. [CrossRef]
35. Demšar, J. Statistical Comparisons of Classifiers over Multiple Data Sets. *J. Mach. Learn. Res.* **2006**, *7*, 1–30.
36. García, S.; Fernández, A.; Luengo, J.; Herrera, F. Advanced Nonparametric Tests for Multiple Comparisons in the Design of Experiments in Computational Intelligence and Data Mining: Experimental Analysis of Power. *Inf. Sci.* **2010**, *180*, 2044–2064. [CrossRef]
37. Friedman, M. The Use of Ranks to Avoid the Assumption of Normality Implicit in the Analysis of Variance. *J. Am. Stat. Assoc.* **1937**, *32*, 675–701. [CrossRef]
38. Wilcox, R.R. *Fundamentals of Modern Statistical Methods: Substantially Improving Power and Accuracy*, 2nd ed.; Springer: New York, NY, USA, 2010; ISBN 978-1-4419-5524-1.
39. Benjamini, Y.; Hochberg, Y. Controlling the False Discovery Rate: A Practical and Powerful Approach to Multiple Testing. *J. R. Stat. Soc. Ser. B Stat. Methodol.* **1995**, *57*, 289–300. [CrossRef]

water

MDPI

Article

Investigation of Local Scouring around Hydrodynamic and Circular Pile Groups under the Influence of River Material Harvesting Pits

Rasoul Daneshfaraz [1,*], Amir Ghaderi [2,*], Maryam Sattariyan [1], Babak Alinejad [1], Mahdi Majedi Asl [1] and Silvia Di Francesco [3]

1 Department of Civil Engineering, Faculty of Engineering, University of Maragheh, Maragheh 8311155181, Iran; maryamsattariyan@gmail.com (M.S.); alinejad@maragheh.ac.ir (B.A.); mehdi.majedi@gmail.com (M.M.A.)
2 Department of Civil Engineering, Faculty of Engineering, University of Zanjan, Zanjan 537138791, Iran
3 Department of Engineering, Niccolò Cusano University, 00166 Rome, Italy; silvia.difrancesco@unicusano.it
* Correspondence: daneshfaraz@maragheh.ac.ir (R.D.); amir_ghaderi@znu.ac.ir (A.G.); Tel.: +98-9144747336 (A.G.)

Abstract: Mining activities can endanger the stability of hydraulic structures. Numerical modeling of local scouring around hydrodynamic and circular bridge pile groups, due to the action of clear water conditions via non-cohesive sediment, was performed using a computational fluid dynamics (CFD) model, a large eddy simulation (LES) turbulence model, and a van Rijn sedimentary model with FLOW-3D software. The pile groups were positioned upstream and downstream of a sand mining pit. The results showed that the scour depth around the downstream pile group was greater than that of the upstream one. Using hydrodynamic piers reduced the scour depth upstream of all piers and the material harvesting pit. The maximum reduction in scour depth was observed in front of the fifth pier, with a 29% reduction in scour depth. Additionally, for all models, as the material harvesting pit was moved downstream, the downstream turbulence was enhanced and stronger flow reversal and horseshoe vortices were detected in from of the downstream pile group. The flow patterns around the pile group showed that the presence of hydrodynamic piers in the upstream pile group leads to a decrease in the maximum flow velocity, whereas, when such piers were positioned in the downstream pile group, the velocity increases.

Keywords: local scouring; bridge pile group; material harvesting pit; hydrodynamic bridge pier; flow pattern; FLOW-3D

Citation: Daneshfaraz, R.; Ghaderi, A.; Sattariyan, M.; Alinejad, B.; Asl, M.M.; Di Francesco, S. Investigation of Local Scouring around Hydrodynamic and Circular Pile Groups under the Influence of River Material Harvesting Pits. *Water* **2021**, *13*, 2192. https://doi.org/10.3390/w13162192

Academic Editor: Vlassios Hrissanthou

Received: 4 June 2021
Accepted: 9 August 2021
Published: 11 August 2021

1. Introduction

Increased transportation requirements have increased the need for wide bridges supported by large piers. Large piers are less hydraulically efficient, since their supports occlude a large flow area and thus disrupt the main flow patterns. As an alternative to singular large piers, it is possible to adopt groups of piers with smaller dimensions.

Furthermore, harvesting of riverbed materials is one of the most common methods used to provide the materials needed in civil works [1]. In fact, sands that are exposed to continuous water flow are a good source of quality granular materials, because their loose materials have been removed by abrasion. Therefore, materials harvested from the riverbed require less processing than other materials, and due to the proximity of rivers to transportation roads and administrative neighborhoods, they reduce excessive transportation costs.

Activities such as harvesting river materials can cause additional concerns. For example, the removal of materials from a river can alter the channel, river hydraulics, or sediment budget, which in turn can alter the distribution of habitats and ecosystem functioning [1]. If harvesting is not performed correctly, disruption can be worsened with consequent long-term damage to the river environment. As examples, the Talar and Balaroud Bridges in Iran are cases where the bridges were destroyed because of scouring resulting from the harvesting of river materials.

Previously, several research efforts have studied scouring around groups of bridge piers, but scouring in the presence of harvesting pits has rarely been investigated. Among the studies of group piers is the study of Ataei-Ashtiani and Beheshti [2], which investigated local scour around pile groups. They observed that the use of pile groups instead of single piers increases the scouring around the piles. Zhou and Alam [3] conducted an extensive review of studies on the flow pattern of dual piers. These researchers shared their findings as a full picture of the flow patterns and discussed the Reynolds number effect, momentum transport characteristics, and fluid forces. Liang et al. [4] performed a laboratory study on scouring around a single pier, two piers next to each other, and a three-on-three pile group. They provided empirical relationships to calculate scour depth. Solaimani et al. [5] studied the effect of the distance and arrangement of bridge piers on the depth, length, and volume of scouring around the piers. They found that increasing the distance between piers increases the depth, area, and the volume of scour. Vaghefi et al. [6] investigated the effect of the vertical trinity of arranged piers on the topographic changes of the bed in a laboratory flume with a 180-degree bend. The maximum scour depth occurred with piers arranged perpendicular to the flow and at an angle of 90 degrees to the bend. Namaee and Sui [7] investigated local scouring around four piers of side-by-side piers under two flow conditions: Open channel and ice-covered flow. They observed, for both cases, that the maximum scour occurred in front of the piers. Moreover, they provided an equation to estimate the scour depth. Malik and Setia [8] conducted a laboratory study of scouring around bridge pile groups in three arrangements: Piles adjacent to each other, aligned in the flow direction, and in a triangular arrangement. For piers arranged in a row, if the distance between the piers exceeds 16 pier diameters, the piers behave as independent entities. Yang et al. [9] investigated the dimensionless ratio of distance to the pier diameter and the Froude number and their effect on scouring. They showed that increasing the ratio of separation distance to the pier diameter leads to a decrease in scour depth. They also found that with an increasing Froude number, there were increases in the dimensionless ratio H_e/D (H_e: scour depth at equilibrium time; D: pier diameter).

There are many studies on the mining of riverbed materials. Decker et al. [10] studied the effects of river material harvesting on the life and environment of fish in the rivers of Georgia in the United States. They concluded that material harvesting has detrimental effects on the environmental conditions of fish. Lee et al. [11] presented regression equations for the migration speed of a rectangular pit. They observed channel degradation in both the upstream and downstream locations of the pit. Cao and Pender [12] provided a 2D numerical study to predict the hydrodynamics of alluvial rivers subjected to interactive sediment mining. The 2D model predicted bed lowering in the downstream side of the mining region. Experiments conducted by Barman et al. [13,14] reported a rise in the turbulent stresses in the mining pit and in the downstream side of the pit. With increasing discharge and length-to-width ratio of the pit, the migration rate of the upstream edge of the pit increases. Extensive studies have been conducted by other researchers about mining pits in a channel, as well as empirical approaches to the migration of pit and channel bed deformation [15–17]. To the best of the authors' knowledge and according to the existing literature from Lee at al. [11] and Barman et al. [13,14], the excavation of a pit in an alluvial channel can induce erosion in the downstream side and can cause streambed instabilities around hydraulic structures such as bridge piers. Lade et al. [18] conducted experimental research on a rectangular pit that was excavated upstream of two circular piers embedded in the sand-bed in a tandem arrangement. They found that mining activities near the piers can lead to significant changes in the flow field, causing excessive scour around piers. Majedi-Asl et al. [19] studied the effect of river material harvesting pits on scouring around pile groups under different hydraulic and sedimentary conditions. They observed that an increase in the Froude number and the presence of a river material harvest pit increases the scour equilibrium time. They also used experimental methods to calculate scour depth and developed a regression method for predicting the depth. Daneshfaraz et al. [20] investigated the effect of flow rate on scour around pile groups bridged with sand in the presence of a material harvesting pit. They concluded that with an increasing flow rate, scour around the base group increased and rough sand on the surface of the foundations reduced scouring around said foundations and increased the scouring time.

A review of previous studies shows that despite the importance of material harvesting pits on local scouring around pile groups, few studies have been conducted in this field. The present study is an attempt to fill these gaps and to provide a better apprehension of these effects. The research is focused on the alterations in the flow parameters and local scouring around three groups of piers placed in the middle of the channel and in arrangements upstream and downstream of an excavated oval pit. Advanced numerical simulation software, FLOW-3D, was used to simulate the problem, and numerical simulations were validated with laboratory experiments. In addition, these works used two shapes of grouped piers in the simulation, namely, circular and hydrodynamic. Streambed alterations, such as scour depth and bed levels and alterations in flow parameters, such as mean velocity around the piers, were investigated.

2. Dimensional Analysis

The scouring around a bridge pile group in the presence of a material harvesting pit is influenced by parameters related to the material harvesting pit, fluid flow, sediment, flume geometry, bridge piers, and scour pit. A functional relationship connecting the dependent and variables with scouring can be provided as follows:

$$f_1(V, y_0, g, \rho, v, d_{50}, \sigma_g, \rho_s, B, D, K_p, K_r, m_p, L_{sx}, L_{sy}, d_s, l_p, b_p, t) = 0 \quad (1)$$

where V is the upstream velocity, y_0 is the initial flow depth, g is the gravitational acceleration, ρ is the fluid density, v is the kinematic viscosity, d_{50} indicates the average particle diameter size, σ_g is the deviation from average size, ρ_s represents the sediment density, B is the flume width, D is the pier diameter, K_p is the curvature of the upstream surface of the pier, K_r is the pier surface roughness, m_p is the length of the pier at the plan, L_{sx} refers to the scour width, L_{sy} is the scour length, d_s indicates the scour depth, l_p represents the length of the harvest pit, b_p is the width of the harvest pit, and t is the time.

Considering ρ, V, and D as iterative variables and using the π-Buckingham method, the dimensionless parameters can be expressed as follows:

$$f_2(\frac{y_0}{D}, \frac{V}{\sqrt{gD}}, \frac{V.D}{v}, \frac{B}{D}, \frac{\rho_s}{\rho}, \frac{d_{50}}{D}, \frac{m_p}{D}, \frac{K_r}{D}, \frac{V.t}{D}, \frac{L_{sx}}{D}, \frac{L_{sy}}{D}, \frac{d_s}{D}, \frac{l_p}{D}, \frac{b_p}{D}, K_p, \sigma_g) = 0 \quad (2)$$

By inserting the dimensionless equivalents ($\frac{V}{\sqrt{gD}} = Fr$, $\frac{V.D}{v} = Re$, $\frac{\rho_s}{\rho} = G_s$) and dividing the dimensionless parameters, we can simplify Equation (2) as follows:

$$f_3\left(Fr, \text{Re}, G_s, \frac{d_{50}}{D}, \frac{m_p}{D}, \frac{K_r}{D}, \frac{V.t}{D}, \frac{L_{sx}}{L_{sy}}, \frac{d_s}{D}, \frac{B}{D}, \frac{l_p}{b_p}, K_p, \sigma_g\right) = 0 \quad (3)$$

According to Mastbergen and van den Berg [21], if $Re \geq 7000$, the kinematic viscosity will not affect local scouring. Considering that in the present study, the Reynolds number for the experiment was 25,600, the Reynolds number (Re) can be ignored. According to Chiew and Melville [22], if $B/D = (1.2/0.09) \geq 10$, the effect of the flume walls on the local scour depth can also be neglected, and consequently, the dimensionless parameter B/D was omitted. Then, due to the uniform distribution of sediment particles and the constancy of the sediment type, their effect on scour depth was neglected [23,24]. The simplified dimensionless relationship can now be expressed as:

$$f_4\left(Fr, \frac{d_{50}}{D}, \frac{m_p}{D}, \frac{K_r}{D}, \frac{V.t}{D}, \frac{L_{sx}}{L_{sy}}, \frac{d_s}{D}, \frac{l_p}{b_p}, K_p\right) = 0 \quad (4)$$

3. Numerical Model

3.1. Flow Field Simulation

FLOW-3D v11.2 software was used to simulate the turbulent three-dimensional flow around the bridge pile group. This program evaluates the location of flow boundaries by utilizing two modeling methods: The FAVOR method (area–volume fraction method), to simulate surfaces and rigid volumes (such as geometric boundaries), and the VOF method (volume of fluid method), to determine the free surface behavior [25]. The ruling equations of the flow include continuity and conservation of momentum. The continuity equation is expressed as follows:

$$V_F\frac{\partial \rho}{\partial t} + \frac{\partial}{\partial x}(\rho u A_x) + \frac{\partial}{\partial y}(\rho v A_y) + \frac{\partial}{\partial z}(\rho w A_z) = R_{SOR} + R_{DIF} \quad (5)$$

In this equation, V_F is the ratio of the volume of fluid passing through an element to the total volume of the element and ρ is the density of the fluid. The parameters u, v, and w are the velocities in the three coordinate directions. The symbols A_x, A_y, and A_z indicate the fraction of area for flow in the x, y, and z directions, respectively. The terms R and x correspond to the selected coordinate system and take on values of 1 and 0, respectively [26].

On the right-hand side of the equation, the term R_{DIF} represents a turbulent diffusion and R_{SOR} is the mass source. It is related to the porous media and the entry of secondary flow into the computational domain. The equations of motion are as follows [27]:

$$\frac{\partial u}{\partial t} + \frac{1}{V_F}\left(uA_x\frac{\partial u}{\partial x} + vA_y\frac{\partial u}{\partial y} + wA_z\frac{\partial u}{\partial z}\right) = -\frac{1}{\rho}\frac{\partial P}{\partial x} + G_x + f_x \quad (6)$$

$$\frac{\partial v}{\partial t} + \frac{1}{V_F}\left(uA_x\frac{\partial v}{\partial x} + vA_y\frac{\partial v}{\partial y} + wA_z\frac{\partial v}{\partial z}\right) = -\frac{1}{\rho}\frac{\partial P}{\partial y} + G_y + f_y \quad (7)$$

$$\frac{\partial w}{\partial t} + \frac{1}{V_F}\left(uA_x\frac{\partial w}{\partial x} + vA_y\frac{\partial w}{\partial y} + wA_z\frac{\partial w}{\partial z}\right) = -\frac{1}{\rho}\frac{\partial P}{\partial z} + G_z + f_z \quad (8)$$

In these equations, P is the pressure; G_x, G_y, and G_z are the accelerations of gravity in the x, y, and z directions; and f_x, f_y, and f_z are the viscous forces, respectively, in the three coordinate directions. The initial conditions for the momentum are as follows:

A. The fluid is continuous and the stress is linearly related to the strain rate;
B. The fluid is isotropic, i.e., the properties of the fluid are independent of direction. As a result, the law of deformation is independent of the chosen axis;
C. When the strain rate is zero, the law of deformation is reduced to hydrostatic pressure.

3.2. Turbulence Model

A turbulence model is needed for additional modeling of the nonlinear Reynolds stress term. Some numerical studies have proposed the Reynolds-Averaged Navier–Stokes (RANS)-based k-omega and the Renormalization Group (RNG) k-ε [28–37] and LES (large eddy simulation) models [38–40] for the scouring problem. The LES turbulence model was used to perform computation of fluid dynamics (CFD). Compared to the RANS models, LES is more computationally demanding. Its application will be more extensive with the development of computer technology. According to Kirkil et al.'s [33] study, the LES turbulence and van Rijn sediment transport models allow to achieve a more accurate result with providing the mesh grids, and are sufficient compared to other models. LES kinematic eddy viscosity (ν_T) performs as:

$$\nu_T = (cL)^2 \sqrt{2e_{ij}2e_{ji}} \tag{9}$$

where c is a constant (0.1–0.2), L is the length scale, and e_{ij} is the strain rate tensor components in the i and j directions.

3.3. Sediment Scour Model

Since many parameters affect scouring, a method based on experimental results can be used to estimate its value. This approach was used in [30,41] to forecast the critical Shields parameter according to the following formulation:

$$T = d\left[\frac{\rho(\rho_s - \rho)g}{\mu^2}\right]^{1/3} \tag{10}$$

where μ is the dynamic viscosity, ρ_s is the fluid density, d is the particle diameter, and g is the acceleration of gravity. Equation (11) is applied to calculate the critical shear stress of sediment in a flat riverbed:

$$\tau_{cr} = \rho g(s-1)d_{50}\theta_{cr} \tag{11}$$

where τ_{cr} is the critical shear stress of the sediment, s is the specific density, d_{50} is the mean diameter of the sediment particles, and θ_{cr} is the critical Shields number. The Soulsby–Whitehouse equation is used to calculate the dimensionless critical Shields parameter:

$$\theta_{cr} = \frac{0.3}{1+1.2T} + 0.055[1 - \exp(-0.02T)] \tag{12}$$

In reality, the bed of a river is not flat, so the critical Shields number can be modified for sloping surfaces to include the angle of repose as follows:

$$\theta_{cr}^{/} = \theta_{cr}\frac{\cos\psi\sin\beta + \sqrt{\cos^2\beta\tan^2\varphi - \sin^2\psi\sin^2\beta}}{\tan\varphi} \tag{13}$$

where β is the angle of the slope of bed, φ is the angle of the repose for sediment (the default is 32°), and ψ is the angle between the flow and the upslope direction. The sediment transfer rate is one of the important criteria for estimating scour; accurate scouring predictions can be achieved if the sediment transport equations are modified. In this study, because of using non-cohesive sediment, the van Rijn equation was used to calculate the dimensionless bed load transfer rate:

$$\phi = \beta_{VR}T\left(\frac{\theta}{\theta_{cr}} - 1\right)^{2.1} C_b \tag{14}$$

where β_{VR} is van Rijn coefficient and is equal to 0.053, C_b is bed load coefficient, and ϕ is the dimensionless bed load transport rate, which is related to the volumetric bed load transport rate, q_b, as follows:

$$q_b = \phi\left[gd^3\left(\frac{\rho_s - \rho}{\rho}\right)\right]^{1/2} \tag{15}$$

This equation computes the bed load transport rate in units of volume per bed width per time. The law of the wall for the mean velocity, as modified for roughness, is expressedby Equation (16):

$$\frac{u_p}{u^*} = \frac{1}{k}\left[E\frac{\rho c_\mu^{0.25}k_p^{0.5}z_p}{\mu}\right] - \Delta B \tag{16}$$

where u_p is the main flow velocity at point p, k is von Kàrmàn's constant (0.418), u^* is the shear velocity related to the bottom shear stress ($\tau = \rho u^{*2}$), E is Young's modulus, ρ is the fluid density, c_μ is the constant, kp is the turbulent kinetic energy at point p, z_p is the distance from point p to the wall, and ΔB is the roughness function [42]. In this study, the parameters selected for sediment scour, obtained after calibration of many runs, were the critical Shields number (θ_{cr}) of 0.03 and the bed load coefficient (C_b) of 0.5

3.4. Description of the Laboratory Experiment and Numerical Setup

This research was carried out in a laboratory flume with a length of 13 m, a width of 1.2 m, and a height of 0.8 m in the hydraulic laboratory of the University of Maragheh, Iran. The flume was constructed using plexiglas walls. A turbulent grid was placed at the inlet to reduce the turbulence of the flow [43]. At the outlet, a metal sluice was installed to adjust the flow depth. Based on insight from Ataei-Ashtiani and Beheshti [2], if x/D (x is the distance between piers and D is the pier diameter) is between 0.15 and 2, the piers behave the same as the pile group. If this ratio becomes less than 0.15, all piers will behave like a large, single pier. However, if the ratio exceeds 2, each pier of the pile group will behave independently. Here, three PVC pipes with a diameter of 0.09 m, positioned at a distance of 0.12 m, were used upstream and downstream of a harvesting pit. In the middle of the flume, a sedimentary bed with a length of 4.26 m, a width of 1.2 m, and a height of 0.226 m was created. According to Raudkivi and Ettema [44], if the D/d_{50} exceeds 20–25, the size of the sediment particles does not affect the final scour depth. They also showed that the average particle diameter must be more than 0.7 mm to prevent the formation of ripples in the sedimentary bed. Therefore, the selected bed sediments were non-cohesive sands with a median size (d_{50}) of 0.86 mm, a specific gravity (G_s) of 2.65, and a geometric standard deviation (σ_g) of 1.32. An oval pit was constructed with large and small dimensions of 1.01 and 0.8 m, respectively. The pit was located in the middle of the bed. Three groups of bridge piers were installed consecutively 1 m apart upstream and downstream of the material harvesting pit.

The physical model was tested under clear water conditions. The flow depth was maintained at 0.132 m. The flow was directed into the flume by a pump capable of providing 7–52 L/s. This flow was sufficient to cause scouring. For experiments, after leveling the sediment bed, it was saturated with a stream of water so that the sediment bed did not change suddenly. Then, the desired hydraulic conditions were adjusted by the gate installed in the flume and water pump, and the test started from that moment. Data were also collected during the experiments. At the end of the tests, the pump was turned off and the flow was slowly drained through the end gate. Details of experiments utilized for validation are shown in Table 1.

Table 1. Specifications of the parameters for validation.

Test No.	Q (L/s)	y_0 (m)	u (m/s)	Fr	Re	u/u_{cr}	d_{50} (m)	G_s	σ_g
T1	45	0.132	0.284	0.25	25,600	0.72	0.00086	2.65	1.32

The data acquired in this study include the scouring depth around the piers at 30 min intervals. Using a 3D laser scanner, water depths were collected with a measurement accuracy of ±1 mm. In both the laboratory experiment and the numerical simulation, all data were gathered from exactly in front of the piers. Figure 1 shows the sediment bed with the relevant details annotated.

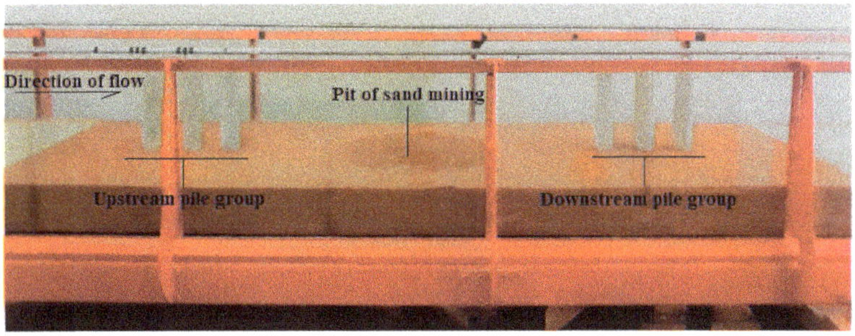

Figure 1. Sediment bed, material harvesting pit, and arrangement of the bridge pile groups in the laboratory model.

The flow geometry, built in AutoCAD software, was then inserted into the computational fluid dynamics program FLOW-3D (see Figure 2). In order to complete the simulation of the flow, the solution region was subdivided into fine cells with a uniform block mesh (Figure 3). The boundary conditions for the numerical model were selected to match the conditions created in the laboratory model. The inlet condition was a uniform velocity, while the output condition was a zero gradient on the transported variables (a continuative boundary condition). According to [44], if the ratio of the flume width to the diameter of the piers (B/D) exceeds 6.25, the effect of the flume walls on scouring can be ignored. Accordingly, if the width of the flume become 0.8 m, the wall effect is ignored, so in order to reduce simulation time, the block mesh's width was assumed to be 0.8 m with symmetry boundary conditions (see Figure 3). "No special conditions are required at the boundaries of symmetry, because there are derivatives of zero velocity at that boundary, and hence the production of zero turbulence. Also, there is a no flow area that does not automatically guarantee any advertising flux or diffusion. Additionally, at these borders, the nodes can slide freely along the border, but they cannot penetrate and retreat."

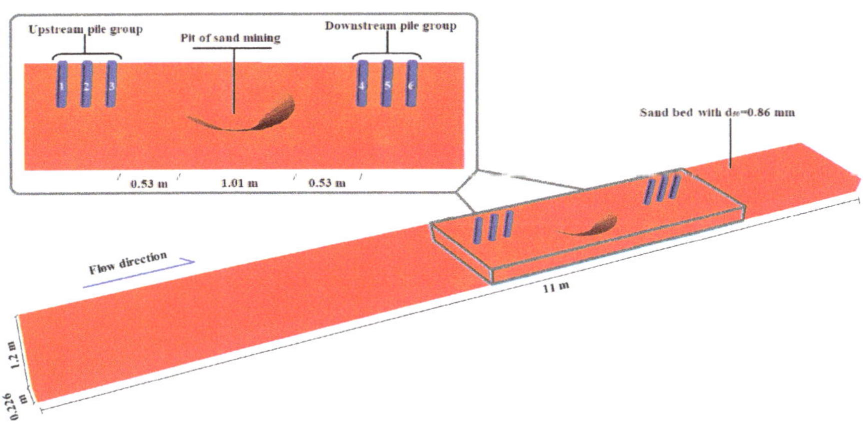

Figure 2. Simulated laboratory flume.

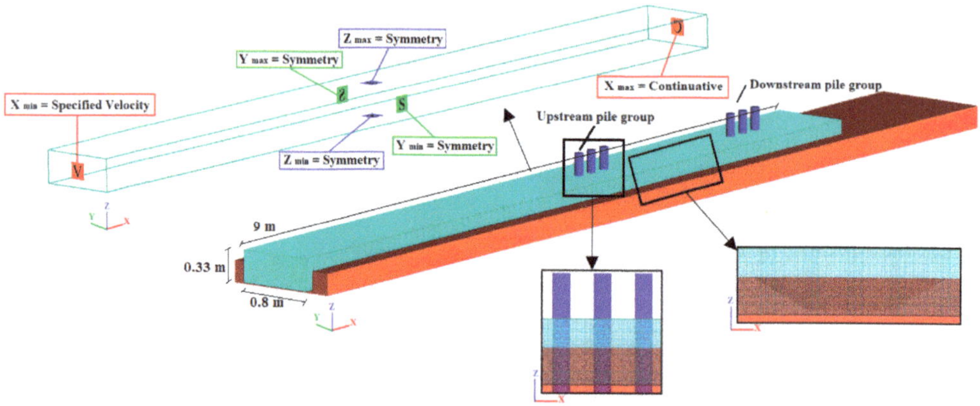

Figure 3. Annotated boundary conditions.

3.5. *Effect of the Computational Mesh on the Scour Depth Results*

Sensitivity of numerical solutions to the meshing is an important consideration in computational studies. To validate the results of the numerical solution, three different computational meshes were constructed and the resulting scour depths are presented in Table 2. In this table, M_{exp} is the experimental value and M_{cal} is the calculated data. From this table, with a decreasing mesh size, the differences between the calculations and measurements decreased. For 2 cm meshes, there was a large difference between the calculations and the measurements (i.e., 84%). However, when the mesh sizes were reduced to 0.9 cm, the difference was reduced to 6.7% for the upstream pier. Similar improvements were observed at the downstream site as well. Table 2 provides a summary list of the results for the three different mesh sizes (0.02, 0.015, and 0.009 m).

Table 2. Comparison of the numerical results for different mesh cell sizes with the laboratory results.

Model	Number of Cells	d_s of the 1st Pier Upstream (m)	$E(\%) = \frac{M_{exp}-M_{cal}}{M_{exp}} \times 100$	d_s of the 1st Pier Downstream (m)	$E(\%) = \frac{M_{exp}-M_{cal}}{M_{exp}} \times 100$
Physical model	-	0.1	-	0.124	-
Mesh size 0.02 m	307,360	0.016	84	0.024	80
Mesh size 0.015 m	707,762	0.047	53	0.056	54
Mesh size 0.009 m	3,332,516	0.107	6.7	0.135	8.8

3.6. *Validation of the Numerical Simulation*

The applicability of a numerical model to realistically reproduce the flow field of a particular application needs to be validated using physical model data (Figure 4a,b). In the laboratory experiments, we reached their equilibrium state in 540 min. The equilibrium conditions correspond to the time after which evolution of the scour hole ceased. Based on the scouring–time chart of the numerical simulation (Figure 4c), the simulation equilibrium was evident, and the modeling time was therefore reduced and set to 360 s. Comparison of the two numerical and laboratory experiments was performed using dimensionless time (Figure 4d). Comparison of the two charts shows good performance of the numerical simulation in simulating the scouring process. In this figure, t_{total} is the total time period of the simulation and $d_{s\ total}$ is the final scour depth in the equilibrium state.

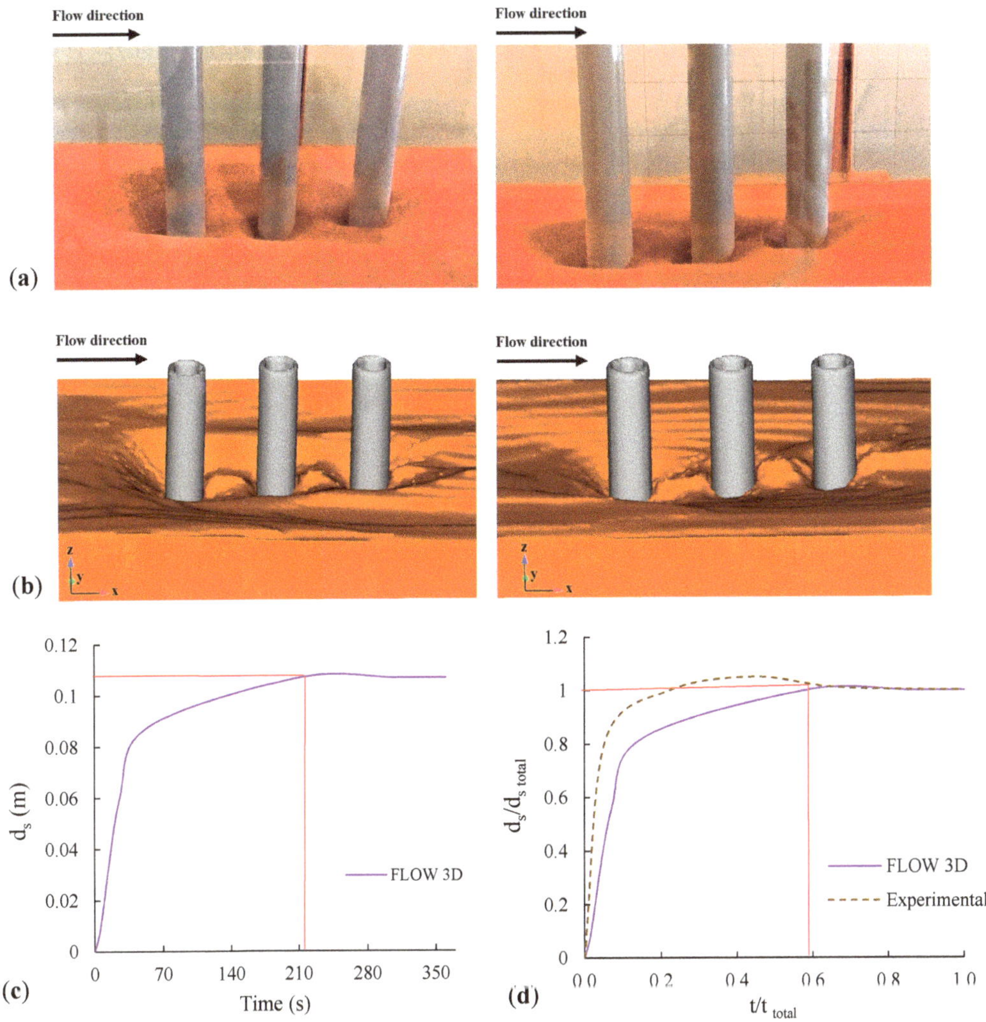

Figure 4. Comparison of scouring around the piers between the laboratory and numerical studies. (**a**) Laboratory scouring around the upstream and downstream pile groups. (**b**) Numerical scouring around the upstream and downstream pile groups. (**c**) Time changes of the local scouring depth in the laboratory model. (**d**) Comparison of the time changes of the local scouring depth between the laboratory model and the numerical simulation.

The validation of the numerical results for the pile group simulations followed a different process than for flow around a single pier. The flow and scouring around the first pier (where maximum scouring occurs) were used as the validation metric. In order to compare the laboratory results with the numerical results, the criteria that were utilized were the relative error percentage (E (%)), the root mean square error ($RMSE$), the coefficient of determination (R^2), and the mean absolute error value (MAE). The definitions of these metrics are provided below:

$$E(\%) = \frac{M_{\text{exp}} - M_{cal}}{M_{\text{exp}}} \times 100 \tag{17}$$

$$RMSE = \sqrt{\frac{1}{N}\sum_{1}^{N}(M_{exp} - M_{cal})^2} \quad (18)$$

$$R^2 = \left(\frac{(N\sum M_{exp}M_{cal}) - (\sum M_{exp})(\sum M_{cal})}{\sqrt{N(\sum M_{exp}^2) - (\sum M_{exp})^2}\sqrt{N(\sum M_{cal}^2) - (\sum M_{cal})^2}}\right)^2 \quad (19)$$

$$MAE = \frac{\sum_1^N |M_{exp} - M_{cal}|}{N} \quad (20)$$

where N is the number of experimental tests, M_{exp} refers to the experimental values, and M_{cal} refers to the calculated data. The fitting equations are acceptable when the RMSE values are close to zero and R^2 is ~1. From Table 3, it can be seen that based on the accuracy of the parameters applied (R^2, MAE, $RMSE$, and E in Table 3), the numerical results are acceptable. For instance, with the first pier (relative error percentage of 6%) and the fourth pier (relative error percentage of 7%), the agreement is excellent. These two piers are the most upstream piers in the two pier groups.

Table 3. Validation of the numerical simulations.

Pier Number	The Dimensionless Scour Depth at the End of the Laboratory Model	The Dimensionless Scour Depth at the End of the Numerical Model	R^2	MAE	RMSE	E
1st	1.1226	1.1876	0.99	0.05	0.06	0.06
2nd	1.0989	0.9404	0.99	0.10	0.13	0.15
3rd	0.7647	0.9826	0.93	0.10	0.14	0.21
4th	1.405	1.5009	0.92	0.14	0.21	0.07
5th	0.9940	1.020	0.95	0.08	0.15	0.03
6th	0.6710	1.0792	0.98	0.14	0.21	0.36

3.7. Simulation of Local Scouring around the Hydrodynamic Pile Group

To create non-circular piers, their cross-sections were modified, as shown in Figure 5. These piers are called hydrodynamic piers, which have sharp noses that coordinate with the flow streamlines and follow the equations of hydrodynamic flow conditions. Based on Ghaderi and Abbasi's [32] study, keen and round nose piers reduce the bed shear stress value more than a circular pier does. According to the recommendations of Ataei-Ashtiani and Beheshti [2], bridge piers with a length of 13.5 cm and a diameter of 9 cm were placed at intervals of 7.6 cm.

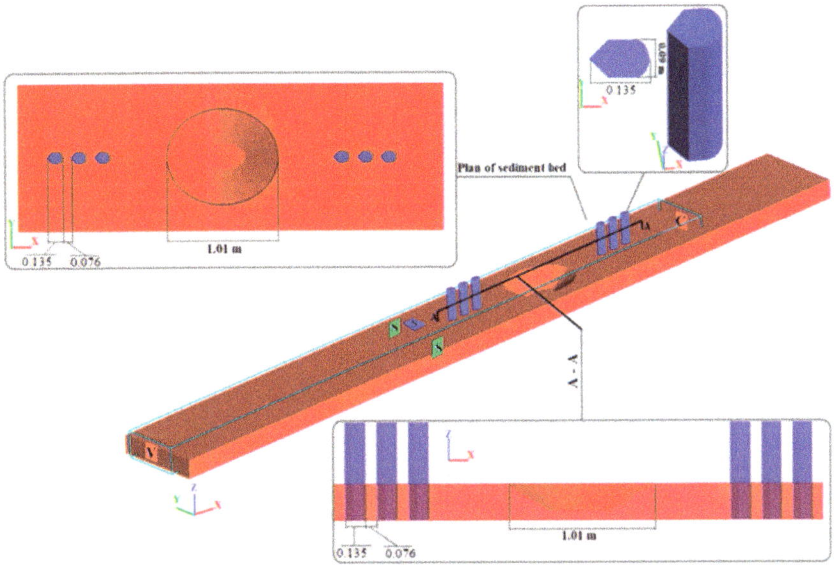

Figure 5. 3D view of the hydrodynamic pile groups and the sediment bed (all dimensions are in meters).

4. Results

First, laboratory-based scouring for circular piers is presented. Then, the scour depth upstream of the hydrodynamic piers is studied and compared with the circular model (both numerical and laboratory results). Finally, by utilizing the computational model, comparisons between circular and hydrodynamic pier scouring are provided.

4.1. Results of Laboratory Study

Figure 6 shows the laboratory results, which include the scour depth around each bridge pier. According to Figure 6, the scour depth around the downstream pile group was greater than that of the upstream pile group. The maximum scour depth around the downstream piers occurred at pier 4 and was 12.4 cm, while for the upstream group, the most significant scouring occurred at pier 1 with a depth of 10.5 cm. The reason for this behavior is the presence of the material harvesting pit, which caused turbulence in the flow and increased the scour depth around the downstream pile group.

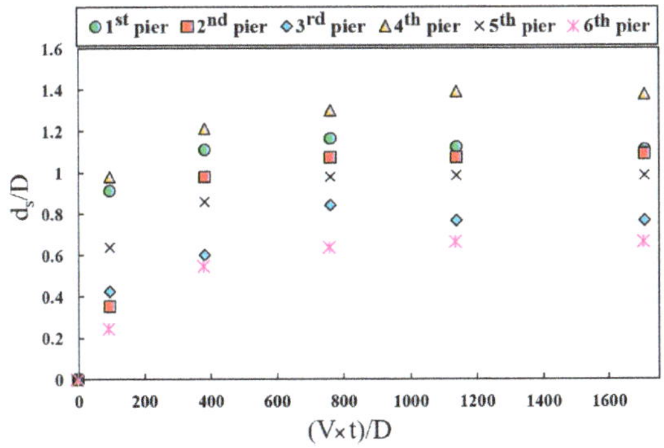

Figure 6. Dimensionless scour depths at various dimensionless times.

4.2. Comparison of Scour for the Hydrodynamic and Circular Pile Groups

Figure 7 shows the evolution of time-dependent scour depths. It can be seen that the use of hydrodynamic piers reduced the scour depths for all piers. According to Figure 7, for the circular model, the maximum scouring depth was 13.4 cm and this occurred at the fourth pier. For the hydrodynamic model, the corresponding maximum scour was, instead, 10.8 cm. A comparison between the circular and hydrodynamic models showed that maximum reduction of the scour depth occurred at the fifth pier in the hydrodynamic model, with a 29% reduction. The reduction in scour depths is related to the reduced turbulence engendered by the hydrodynamic shape. Reducing the turbulence of the flow around the piers reduces the strength of the horseshoe vortices and causes less sediment to be transported downstream by the flow vortices. The simulated maximum scour depths are summarized in Figure 8 and Table 4.

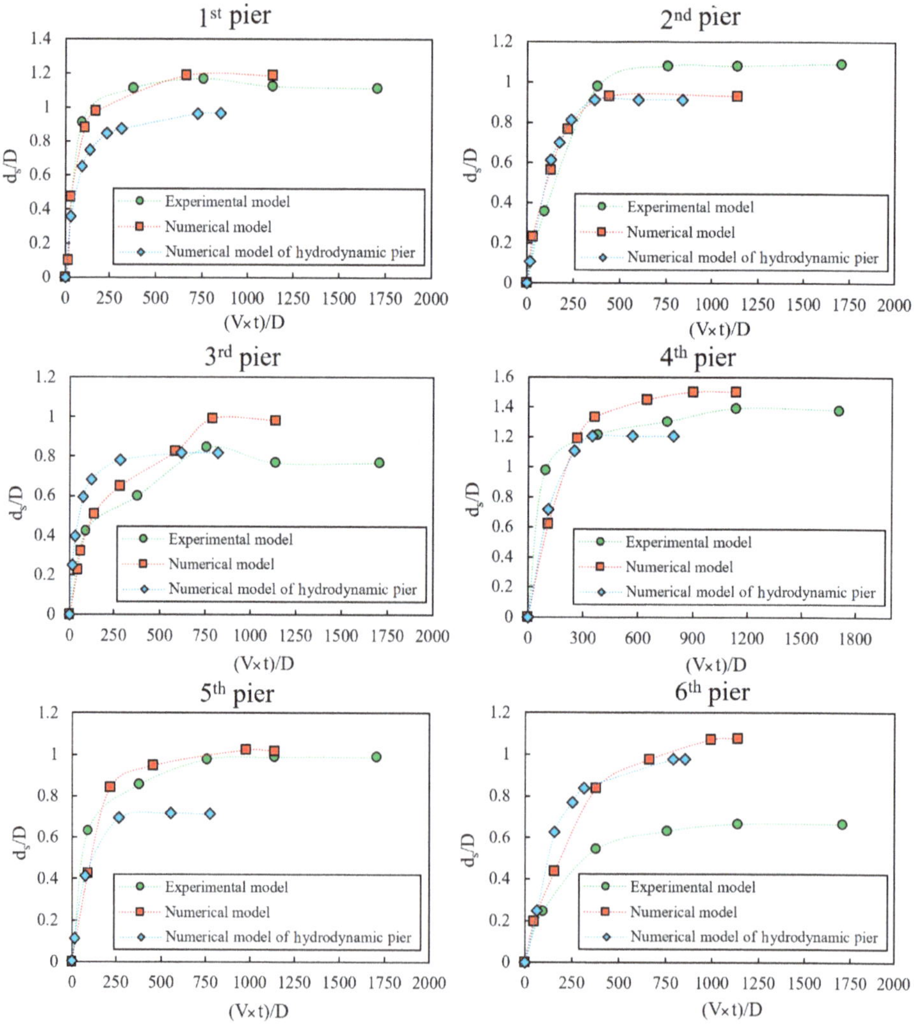

Figure 7. Scour depths for the circular and hydrodynamic piers.

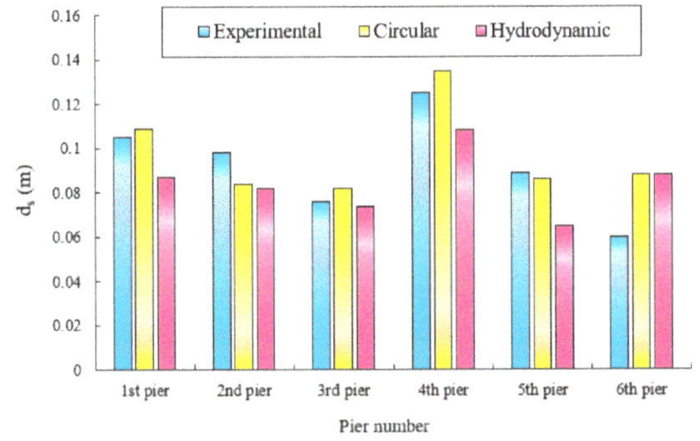

Figure 8. Maximum scour depth in the group piers.

Table 4. Comparison of the maximum scour depths.

Pier Model	1st Pier	2nd Pier	3rd Pier	4th Pier	5th Pier	6th Pier
Scour depth around the hydrodynamic piers (m)	0.0868	0.0821	0.0734	0.1083	0.0643	0.0880
Scour depth around the circular piers (m)	0.1067	0.0838	0.0883	0.1349	0.0915	0.0968

4.3. Flow Patterns around the Hydrodynamic and Circular Pile Groups

The flow patterns around the pile groups were different compared to the single pier. For a group, the first pier was a barrier to the flow and caused turbulence and a vortex flow pattern. Subsequently, the presence of the second pier reduced this turbulence and caused a sheltering effect. The third pier, in addition to playing a protective role for the rearward side of the second pier, also experienced less scouring due to the deposition of washed particles from around the second pier. Figure 9 shows the flow pattern around the pile groups upstream and downstream of the material harvest pit. The graphs correspond to the instantaneous velocity at the end of the simulation, at which point the scouring was balanced. According to this figure, for both the circular and hydrodynamic piers, the flow velocity in the wake regions behind the piers was very small, while a maximum value occurred at the sides of the piers. In the circular model, the flow lines struck the surface of the first pier and then abruptly deflected to the sides. However, for the hydrodynamic piers, the fluid flow was split more gradually, so that sudden flow deviation was avoided.

The presence of the harvest pit caused turbulence in the downstream areas. For the upstream pile group, the prevailing flow was uniform. This uniform flow was disturbed by its collision with the first pier, while the turbulence of the flow around the downstream pile group was due to both collisions of the flow with the downstream pile group and advection of the turbulence from the upstream locations. The hydrodynamic piers led to a reduction in the maximum velocity adjacent to the upstream group and a reduction in the maximum velocity of the downstream group.

For the upstream pile group, the maximum local velocity was associated with a circular shape and took on a value of 0.67 m/s. For the hydrodynamic model, this quantity decreased by 5% to a value of 0.63 m/s. In the downstream pile group, the maximum value occurred with the hydrodynamic model and was 0.67 m/s. Considering that the distance between piers in the hydrodynamic model (0.076 m) was less than in the circular model (0.12), based on Figure 9, it can be found that the protective role between the piers due to increased pier interference in the hydrodynamic model was larger than in the circular model.

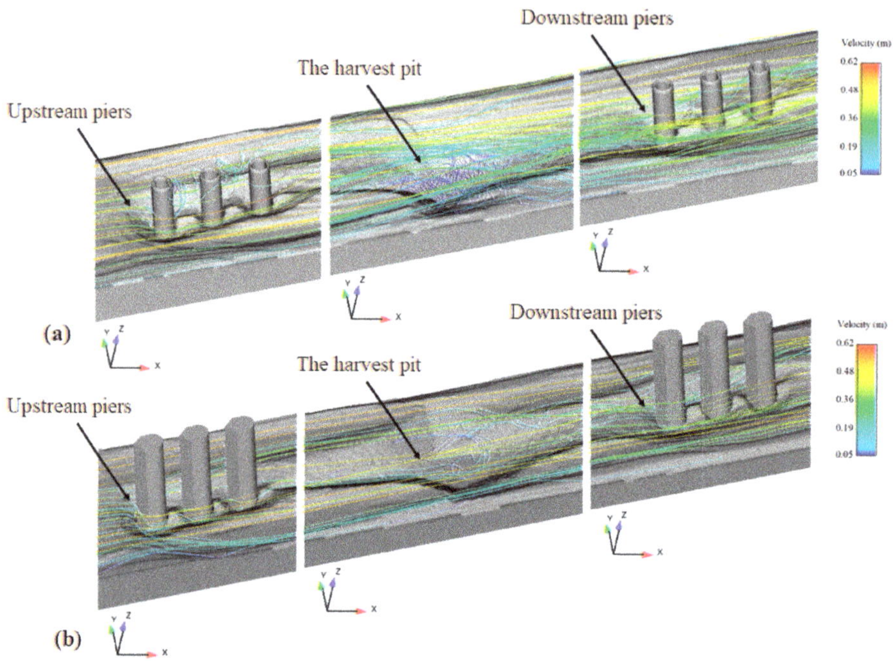

Figure 9. Flow patterns for the pier group and the material harvest pit: (**a**) Circular model; (**b**) hydrodynamic model.

4.4. Investigation of the Material Harvest Pit Changes

The final shapes of the scouring around the group piles and the material harvest pit are shown in Figure 10. In these figures, the negative and positive elevation changes reflect areas that were eroded and built up, respectively. In Figure 10, the upstream portion of the harvesting pit experienced sediment deposition, whereas the more downstream locations experienced erosion. The source of the upstream deposition came from locations such as the crown of the pit and the upstream piles. The harvest pit for the hydrodynamic model exhibited less changes compared to the circular model. The maximum erosion for the circular model was 11.3 cm, which in the hydrodynamic model decreased by 17% to 9.3 cm. The reduction in scouring was related to the reduced volume of particles that had been transferred into the harvest pit (see Figure 11).

Figure 12 shows the longitudinal profiles of the harvest pit in the x–z plane, while Figure 13 shows the changes in the material harvest pit in the y–z plane. The water flow expanded the dimensions of the material harvesting pit in both the transverse and longitudinal directions. However, the expansion of the pit in the downstream direction, relative to the width of the pit, was so great that it can be clearly seen that the material harvesting pit moved downstream.

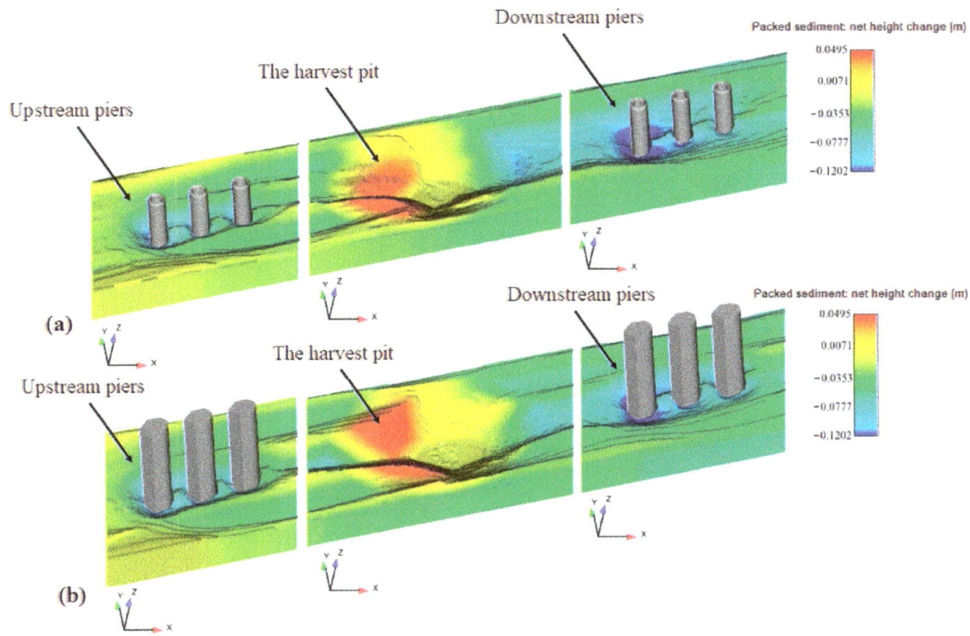

Figure 10. Bed elevation changes from around the group piles and the material harvest pit: (**a**) Circular model; (**b**) hydrodynamic model.

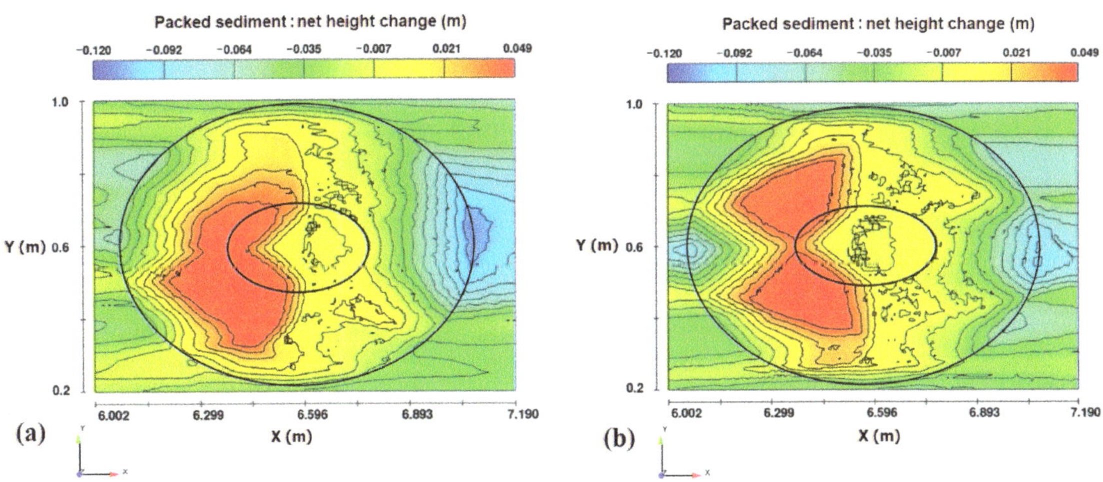

Figure 11. Changes in the harvest pit in the plane (plane x–y): (**a**) circular model, (**b**) hydrodynamic model.

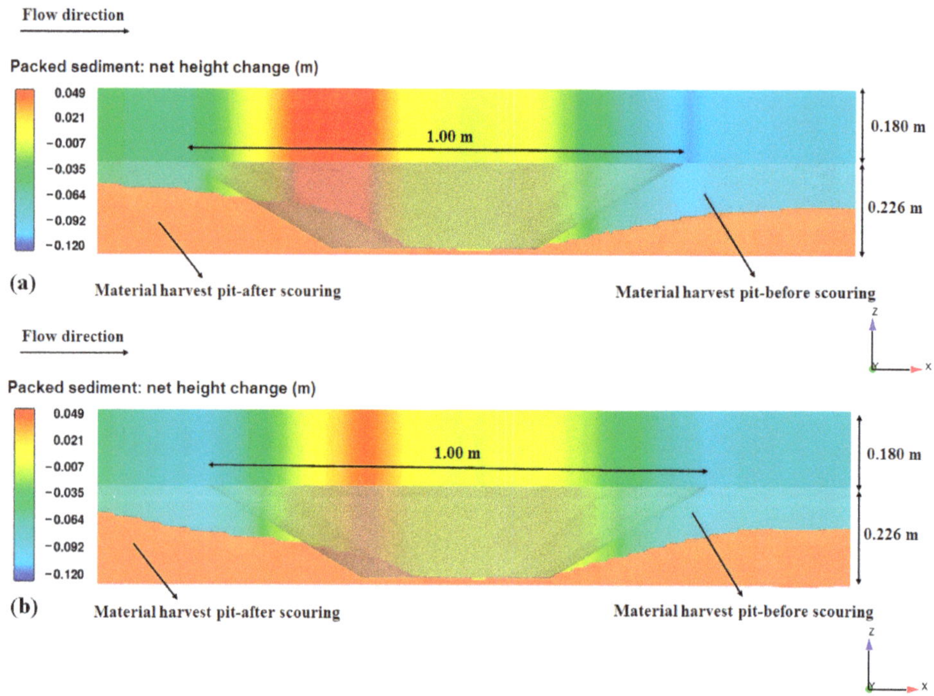

Figure 12. Material harvest pit changes in the x–z plane: (**a**) Circular model; (**b**) hydrodynamic model.

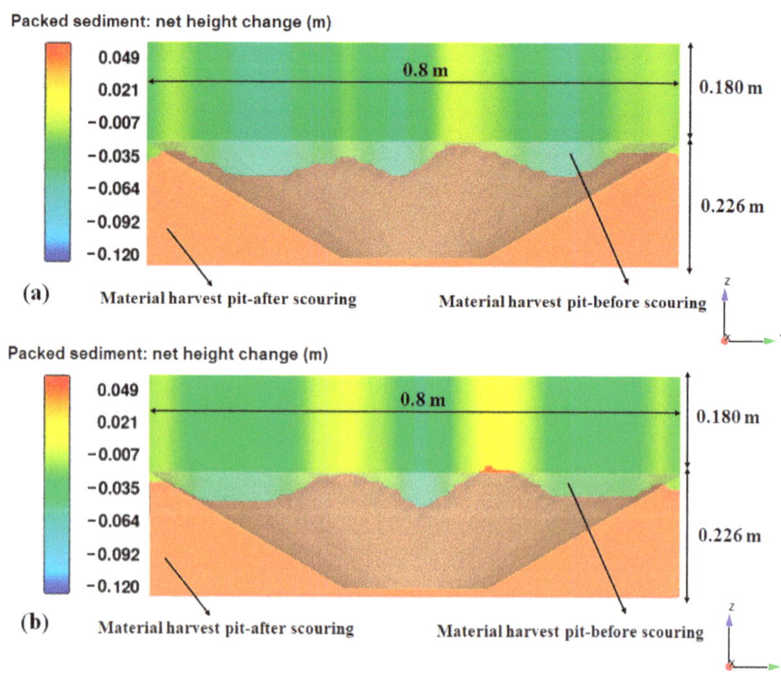

Figure 13. Changes in the harvest pit in the y–z plane: (**a**) Circular model; (**b**) hydrodynamic model.

5. Conclusions

The current study investigated local scour around pile groups positioned upstream and downstream of a harvest pit. Both circular and hydrodynamic shaped piers were considered. The key findings of the comparative analysis are given below.

After passing over a pit, the flow approached the downstream pile group with higher mean velocities and increased turbulent stresses. Thus, the scour depth around the downstream pile group was greater than that of the upstream pile group.

The hydrodynamic piers reduced the scour depth in front of all piers, thanks to the alignment of the piers in the flow and their consequent reduction in turbulence. The maximum scour depth reduction (reduction of 29%) occurred at the fifth pier in the hydrodynamic model.

The material harvest pit was moved downstream and the evolution of local scour was remarkably faster in the presence of a pit, especially in the downstream pile group. However, the changes to the pit were smaller when hydrodynamic piers were used, compared to circular piers. The maximum erosion for the circular piers was 11.3 cm, while for the hydrodynamic piers, the maximum erosion was 9.3 cm (a 17% reduction).

By comparing the flow patterns around the pile groups, both upstream and downstream of the material harvesting pit, it was observed that the presence of the pit caused turbulence in the downstream areas. The evaluation of the flow patterns around the pile groups showed that for both pier shapes, the velocity downstream of the piers was minimal, while the velocities at the sides of the piers were maximum.

Thus, mining activities upstream of pile groups cause local scouring around piers and excess riverbed erosion. These alterations in the flow field and morphology around piers caused by mining activities can jeopardize the stability of such hydraulic structures.

Author Contributions: Conceptualization, R.D., A.G., M.S., B.A., M.M.A. and S.D.F.; methodology, R.D., A.G., M.S., B.A., M.M.A. and S.D.F.; software, A.G. and M.S.; validation, A.G. and M.S.; formal analysis, R.D., A.G. and M.S.; investigation, R.D., A.G., M.S., B.A., M.M.A. and S.D.F.; resources, R.D., A.G. and M.S.; data curation, R.D., A.G. and M.S.; writing—original draft preparation, R.D., A.G., M.S., B.A., M.M.A. and S.D.F.; writing—review and editing, R.D., A.G., M.S., B.A., M.M.A. and S.D.F.; visualization, A.G. and M.S.; supervision, R.D., B.A. and M.M.A.; project administration, R.D., A.G., B.A. and M.M.A. All authors have read and agreed to the published version of the manuscript.

Funding: This research received no external funding.

Institutional Review Board Statement: Not applicable.

Informed Consent Statement: Not applicable.

Data Availability Statement: All data are contained within the article.

Acknowledgments: The authors express their sincere thanks to the Editor, Academic Editor, Associate Editor, and anonymous reviewers for devoting their precious time and providing useful suggestions and constructive comments.

Conflicts of Interest: The authors declare no conflict of interest.

References

1. Koehnken, L.; Rintoul, M.S.; Goichot, M.; Tickner, D.; Loftus, A.C.; Acreman, M.C. Impacts of riverine sand mining on freshwater ecosystems: A review of the scientific evidence and guidance for future research. *River Res. Appl.* **2020**, *36*, 362–370. [CrossRef]
2. Ataei-Ashtiani, B.; Beheshti, A.A. Experimental investigation of clear-water local scour at pile groups. *J. Hydraul. Eng.* **2006**, *10*, 1100–1104. [CrossRef]
3. Zhou, Y.; Mahbub, A.M. Wake of two interacting circular cylinders: A review. *Int. J. Heat Fluid Flow* **2016**, *62*, 510–537. [CrossRef]
4. Liang, F.; Wang, C.; Huang, M.; Wang, Y. Experimental observations and evaluations of formulae for local scour at pile groups in steady currents. *Mar. Georesources Geotechnol.* **2017**, *35*, 245–255. [CrossRef]
5. Solaimani, N.; Amini, A.; Banejad, H.; Taherei Ghazvinei, P. The effect of pile spacing and arrangement on bed formation and scour hole dimensions in pile groups. *Int. J. River Basin Manag.* **2017**, *15*, 219–225. [CrossRef]
6. Vaghefi, M.; Motlagh, M.J.T.N.; Hashemi, S.S.; Moradi, S. Experimental study of bed topography variations due to placement of a triad series of vertical piers at different positions in a 180 bend. *Arab. J. Geosci.* **2018**, *11*, 102. [CrossRef]

7. Namaee, M.R.; Sui, J. Local scour around two side-by-side cylindrical bridge piers under ice-covered conditions. *Int. J. Sediment Res.* **2019**, *34*, 355–367. [CrossRef]
8. Malik, R.; Setia, B. Interference between pier models and its effects on scour depth. *SN Appl. Sci.* **2020**, *2*, 68. [CrossRef]
9. Yang, Y.; Qi, M.; Wang, X.; Li, J. Experimental study of scour around pile groups in steady flows. *Ocean. Eng.* **2020**, *195*, 106651. [CrossRef]
10. Decker, C.; Keyes, J.; Jackson, C.R.; Shelton, J.; Jackson, B. *The Effects of Sand Dredging on Channel Morphology, Habitat, and Water Quality in Urban DeKalb County Streams*; Georgia Institute of Technology: Atlanta, GA, USA, 1999.
11. Lee, H.Y.; Fu, D.T.; Song, M.H. Migration of rectangular mining pit composed of uniform sediment. *J. Hydraul. Eng.* **1993**, *119*, 64–80. [CrossRef]
12. Cao, Z.; Pender, G. Numerical modeling of alluvial rivers subjected to interactive sediment mining and feeding. *Adv. Water Resour.* **2004**, *27*, 533–546. [CrossRef]
13. Barman, B.; Kumar, B.; Sarma, A.K. Turbulent flow structures and geomorphic characteristics of a mining affected alluvial channel. *Earth Surf. Proces. Landf.* **2018**, *43*, 1811–1824. [CrossRef]
14. Barman, B.; Kumar, B.; Sarma, A.K. Dynamic characterization of the migration of a mining pit in an alluvial channel. *Int. J. Sediment Res.* **2019**, *34*, 155–165. [CrossRef]
15. Barman, B.; Sharma, A.; Kumar, B.; Sarma, A.K. Multiscale characterization of migrating sand wave in mining induced alluvial channel. *Ecol. Eng.* **2017**, *102*, 199–206. [CrossRef]
16. Neyshabouri, S.A.A.S.; Farhadzadeh, A.; Amini, A. Experimental and field study of mining pit migration. *Int. J. Sedim. Res.* **2002**, *17*, 323–333.
17. Barman, B.; Kumar, B.; Sarma, A.K. Impact of sand mining on alluvial channel flow characteristics. *Ecol. Eng.* **2019**, *135*, 36–44. [CrossRef]
18. Lade, A.D.; Deshpande, V.; Kumar, B.; Oliveto, G. On the Morphodynamic Alterations around Bridge Piers under the Influence of Instream Mining. *Water* **2019**, *11*, 1676. [CrossRef]
19. Majedi-Asl, M.; Daneshfaraz, R.; Abraham, J.; Valizadeh, S. Effects of Hydraulic Characteristics, Sedimentary Parameters, and Mining of Bed Material on Scour Depth of Bridge Pier Groups. *J. Perform. Constr. Facil.* **2020**, *35*, 04020148. [CrossRef]
20. Daneshfaraz, R.; Sattariyan Karajabad, M.; Alinejad, B.; Majedi-Asl, M. Experimental Investigation of the Effects of Flow Discharge on the Scour Rate around the Groups of Bridge Piers with a Rough Surface in the Presence of Aggregate Extraction Pits. *JWSS* **2021**, *24*, 111–125.
21. Mastbergen, D.R.; van den Berg, J.H. Breaching in fine sands and the generation of sustained turbidity currents in submarine canyons. *Sedimentology* **2003**, *50*, 625–637. [CrossRef]
22. Chiew, Y.M.; Melville, B.W. Local scour around bridge piers. *J. Hydraul. Res.* **1987**, *25*, 15–26. [CrossRef]
23. Melville, B.W.; Chiew, Y.M. Time scale for local scour at bridge piers. *J. Hydraul. Eng.* **1999**, *125*, 59–65. [CrossRef]
24. Ghaderi, A.; Daneshfaraz, R.; Torabi, M.; Abraham, J.; Azamathulla, H.M. Experimental investigation on effective scouring parameters downstream from stepped spillways. *Water Supply* **2020**, *20*, 1988–1998. [CrossRef]
25. Flow Science Inc. *FLOW-3D V 11.2 User's Manual*; Flow Science, Inc.: Santa Fe, NM, USA, 2016.
26. Abbasi, S.; Fatemi, S.; Ghaderi, A.; Di Francesco, S. The Effect of Geometric Parameters of the Antivortex on a Triangular Labyrinth Side Weir. *Water* **2021**, *13*, 14. [CrossRef]
27. Daneshfaraz, R.; Ghaderi, A.; Akhtari, A.; Di Francesco, S. On the Effect of Block Roughness in Ogee Spillways with Flip Buckets. *Fluids* **2020**, *5*, 182. [CrossRef]
28. Baykal, C.; Sumer, B.M.; Fuhrman, D.R.; Jacobsen, N.G.; Fredsøe, J. Numerical simulation of scour and backfilling processes around a circular pile in waves. *Coast. Eng.* **2017**, *122*, 87–107. [CrossRef]
29. Li, Y.; Ong, M.C.; Fuhrman, D.R.; Larsen, B.E. Numerical investigation of wave-pluscurrent induced scour beneath two submarine pipelines in tandem. *Coast. Eng.* **2020**, *156*, 103619. [CrossRef]
30. Ghaderi, A.; Abbasi, S. CFD simulation of local scouring around airfoil-shaped bridge piers with and without collar. *Sādhanā* **2019**, *44*, 216. [CrossRef]
31. Li, Y.; Ong, M.C.; Fuhrman, D.R. CFD investigations of scour beneath a submarine pipeline with the effect of upward seepage. *Coast. Eng.* **2020**, *156*, 103624. [CrossRef]
32. Ghaderi, A.; Dasineh, M.; Aristodemo, F.; Aricò, C. Numerical Simulations of the Flow Field of a Submerged Hydraulic Jump over Triangular Macroroughnesses. *Water* **2021**, *13*, 674. [CrossRef]
33. Fuhrman, D.R.; Schløer, S.; Sterner, J. RANS-based simulation of turbulent wave boundary layer and sheet-flow sediment transport processes. *Coast. Eng.* **2013**, *73*, 151–166. [CrossRef]
34. Choufu, L.; Abbasi, S.; Pourshahbaz, H.; Taghvaei, P.; Tfwala, S. Investigation of Flow, Erosion, and Sedimentation Pattern around Varied Groynes under Different Hydraulic and Geometric Conditions: A Numerical Study. *Water* **2019**, *11*, 235. [CrossRef]
35. Ghaderi, A.; Daneshfaraz, R.; Dasineh, M.; Di Francesco, S. Energy Dissipation and Hydraulics of Flow over Trapezoidal–Triangular Labyrinth Weirs. *Water* **2020**, *12*, 1992. [CrossRef]
36. Ghaderi, A.; Daneshfaraz, R.; Dasineh, M. Evaluation and prediction of the scour depth of bridge foundations with HEC-RAS numerical model and empirical equations (Case Study: Bridge of Simineh Rood Miandoab, Iran). *Eng. J.* **2019**, *23*, 279–295. [CrossRef]

37. Daneshfaraz, R.; Aminvash, E.; Ghaderi, A.; Kuriqi, A.; Abraham, J. Three-Dimensional Investigation of Hydraulic Properties of Vertical Drop in the Presence of Step and Grid Dissipators. *Symmetry* **2021**, *13*, 895. [CrossRef]
38. Kirkil, G.; Constantinescu, G.; Ettema, R. Detached eddy simulation investigation of turbulence at a circular pier with scour hole. *J. Hydraul. Eng.* **2009**, *135*, 888–901. [CrossRef]
39. McCoy, A.; Constantinescu, G.; Weber, L.J. Numerical investigation of flow hydrodynamics in a channel with a series of groynes. *J. Hydraul. Eng.* **2008**, *134*, 157–172. [CrossRef]
40. Koken, M.; Constantinescu, G. An investigation of the flow and scour mechanisms around isolated spur dikes in a shallow open channel: 1. Conditions corresponding to the initiation of the erosion and deposition process. *Water Resour. Res.* **2008**, *44*. [CrossRef]
41. Soulsby, R. Chapter 9: Bedload transport. In *Dynamics of Marine Sands*; Thomas Telford Publications: London, UK, 1997.
42. Omara, H.; Elsayed, S.M.; Abdeelaal, G.M.; Abd-Elhamid, H.F.; Tawfik, A. Hydromorphological numerical model of the local scour process around bridge piers. *Arab. J. Sci. Eng.* **2019**, *44*, 4183–4199. [CrossRef]
43. Ghaderi, A.; Abbasi, S. Experimental and Numerical Study of the Effects of Geometric Appendance Elements on Energy Dissipation over Stepped Spillway. *Water* **2021**, *13*, 957. [CrossRef]
44. Raudkivi, A.J.; Ettema, R. Clear-water scour at cylindrical piers. *J. Hydraul. Eng.* **1983**, *109*, 338–350. [CrossRef]

water

MDPI

Article

Sediment Transport and Water Flow Resistance in Alluvial River Channels: Modified Model of Transport of Non-Uniform Grain-Size Sediments

Gennady Gladkov [1], Michał Habel [2,*], Zygmunt Babiński [2] and Pakhom Belyakov [1]

1 Institute of Water Transport, Admiral Makarov State University of Maritime and Inland Shipping, 5/7 Dvinskaya St., 198035 St. Petersburg, Russia; gladkovgl@gumrf.ru (G.G.); belyakovpv@gumrf.ru (P.B.)
2 Institute of Geography, Kazimierz Wielki University, 85-033 Bydgoszcz, Poland; zygmunt.babinski@gmail.com
* Correspondence: hydro.habel@ukw.edu.pl; Tel.: +48-535-105-104

Abstract: The paper presents recommendations for using the results obtained in sediment transport simulation and modeling of channel deformations in rivers. This work relates to the issues of empirical modeling of the water flow characteristics in natural riverbeds with a movable bottom (alluvial channels) which are extremely complex. The study shows that in the simulation of sediment transport and calculation of channel deformations in the rivers, it is expedient to use the calculation dependences of Chézy's coefficient for assessing the roughness of the bottom sediment mixture, or the dependences of the form based on the field investigation data. Three models are most commonly used and based on the original formulas of Meyer-Peter and Müller (1948), Einstein (1950) and van Rijn (1984). This work deals with assessing the hydraulic resistance of the channel and improving the river sediment transport model in a simulation of riverbed transformation on the basis of previous research to verify it based on 296 field measurements on the Central-East European lowland rivers. The performed test calculations show that the modified van Rijn formula gives the best results from all the considered variants.

Keywords: river sediment transport modeling; alluvial channels; hydraulic riverbed resistance; mathematical modeling; van Rijn sediment transport formula

Citation: Gladkov, G.; Habel, M.; Babiński, Z.; Belyakov, P. Sediment Transport and Water Flow Resistance in Alluvial River Channels: Modified Model of Transport of Non-Uniform Grain-Size Sediments. *Water* **2021**, *13*, 2038. https://doi.org/10.3390/w13152038

Academic Editors: Vlassios Hrissanthou and Bommanna Krishnappan

Received: 29 April 2021
Accepted: 22 July 2021
Published: 26 July 2021

1. Introduction

The problem of forecasting channel transformations in riverbeds with a movable bottom (alluvial channels) is one of the most complex in fluvial hydraulics [1]. Along with traditional research techniques, the methods of mathematical modeling have been and continue to be widely used in the modern practice of hydraulic calculations. In actual practice, the task is to perform hydraulic calculations of the water flow characteristics and sediment transport parameters (bottom deformations) in rivers. In terms of calculation, this boils down to solving by numerical methods a known set of equations of water movement, continuity and deformations at the given initial and boundary conditions. This set of equations in five unknowns is closed by two additional calculation dependences—the hydraulic resistance law in the form of Chézy's formula and sediment discharge formula. Therefore, the quality of channel predictions largely depends on reliability of the energy loss estimates over the length and sediment transport parameters in the deformable channel.

1.1. Hydraulic Resistance Factors to Water Flow in Rivers

Hydraulic resistance of natural channels is one of the biggest challenges of river flow dynamics. The current level of this discipline [2] requires treating the river flow and movable channel as a single system, the interaction of which is based on the principle of feedback (flow ↔ channel). As this takes place, the flow itself creates and regulates the roughness of its bottom. The complexity and lack of knowledge of this issue significantly limits the theoretical approach to its solution. Therefore, the results obtained in recent years are mainly empirical or semi-empirical. In the practice of engineering calculations, either hydraulic friction coefficient λ, or Chézy's formula $C = \sqrt{2g/\lambda}$ are used.

The values of Chézy's coefficient for natural channels are usually derived from the Manning formula [3], i.e., via dimension roughness factor n. Factor n expresses the cumulative effect of all resistances found in the natural channel, and therefore varies considerably. When dealing with long estimated areas, with resistance variations smoothed along their length, it is permissible to use the n values from tables of hydraulic reference guides. When performing calculations on relatively short distances, the use of tabulated values of the roughness coefficients is unacceptable. In this case, n is determined from the measurements of water discharge and slope, which, in turn, requires observations on the survey plot.

In determining the hydraulic resistance in a channel with sandy bottom sediments, another challenge is the relation of its value with the flow rate. Therefore, the use of the Manning formula or similar formulas with tabulated values of the roughness coefficient can result in gross errors for the movable channel.

As for the open channel, Meyer-Peter and Müller (hereafter MPM) suggested presenting the friction slope as the sum of the slope due to grain roughness of the bottom and the slope caused by the bed forms resistance [4]:

$$I = I_d + I_r \tag{1}$$

where I_d is the slope due to grain roughness of the bottom, I_r is the slope caused by the bed forms resistance.

Equation (1) corresponds to a known relationship between the Chézy coefficients [3]:

$$1/C^2 = 1/(C_d)^2 + 1/(C_r)^2 \tag{2}$$

The contribution of each component of the Chézy coefficient C_d and C_r may be different in the different rivers depending on the size of the particles forming the bottom. In rivers with a sand channel, the main part of the lengthwise energy loss is due to the bed form resistance. Research results given by Grishanin [5] show that in the range of values of Chézy's coefficient $C < 50$ m$^{1/2}$/s, which applies to most of the plain rivers with sandy bottom sediments, over 90% of the lengthwise energy loss is due to resistance of the bed form bottom relief.

The situation is different with resistance of the channels composed of coarse sediment. Its value in rivers with gravel and pebble bottom sediments is mainly determined by the grain roughness of the bottom. It depends on the size of the particles of the bottom material, represented as the *H/d* parameter, where H (m) is mean depth in river cross-section = ω/B; ω (m^2) is cross-sectional area; B (m) is cross-section width; d (m) is median sediment diameter.

1.2. Grain Roughness

To estimate the grain roughness, the power formulas with the structure of Manning–Strickler [6] dependence and logarithmic formulas by the type of Zegzhda formula [7] are usually used in the engineering calculations. The Manning–Strickler empirical calculation dependence reads as:

$$C_d/\sqrt{g} = 6.67(R/d_{50})^{1/6} \tag{3}$$

where C_d (m$^{1/2}$/s) is Chézy's coefficient due to grain roughness, g (m/s^2) the gravity acceleration, R (m) is hydraulic radius, and d_{50} (m) is the value of the particle diameter at 50% in the cumulative distribution.

Through experiments in the open trays with pasted roughness, Zegzhda [7] obtained the calculation dependence reads as:

$$C_d/\sqrt{g} = 5.66lg(R/D) + 6.01 \tag{4}$$

where g (m/s^2) the gravity acceleration, R (m) is hydraulic radius, and D (m) is the height of the roughness protrusions, taken, respectively, equal to $D = 1.6d_{50}$ for sandy particles, $D = 1.3d_{50}$ for fine gravel particles, and $D = d_{50}$ for medium and coarse gravel.

The pebble riverbed resistance study was carried out by Griffiths [8]. The initial data in his study were 136 measurements at 72 reaches of 46 New Zealand rivers. Taking into account measurement data of some other authors (a total of 186 points), Griffiths obtained the resistance formula for a channel with an immovable bed:

$$C_d/\sqrt{g} = 5.60lg(R/d_{50}) + 2.15 \quad (5)$$

The analysis of research published on this work shows that the formula of form (4) has been extensively verified based on the field measurements and laboratory data and is recommended for use in calculations by a number of other authors [9–14]. These results indicate the increasing hydraulic resistance when passing from experiments in the trays to the river original flows. This becomes evident in the fact that all the authors have obtained different intersection values in Formula (4) for the constant term K. They vary from $K = 6.01$ of Zegzhda by measurements in trays with the attached sand roughness [7], and $K = 2.15$ [8] by Griffith's Formula (5) to $K = -5.3$ of Grishanin [14] by measurements in the rivers with a pebble-boulder bed. Such significant variations in the values of the constant term K of different authors are due to the heterogeneity of sediment of the bottom sediments and additional resistance factors in the natural channel flows.

The first component of additional energy losses can be considered to some extent due to a more detailed evaluation of the distribution of the coarseness of granulometric composition when selecting the representative diameter of the bottom sediments. Ribberink [15] uses in his study the log normal particle size distribution law for the river alluvium. Considering the results of studies [16,17], the calculation expression for determining the representative diameter value d_a can be represented, which reads as:

$$d_a = 1.6d_{50}\sigma_g^{-0.56} \quad (6)$$

where $\sigma_g = \exp\sqrt{ln\left[(S_{dm}/d_m)^2 + 1\right]}$ is the geometric standard deviation of the diameter logarithm from the mean by the value equal to the standard deviation; otherwise—$\sigma_g = 0.5[d_{84}/d_{50} + d_{50}/d_{16}]$, where d_{84} and d_{16} are characteristic particle diameters equal to the 84- and 16-percentile size, respectively.

$S_{dm} = \sqrt{\sum\beta_i(d_i - d_m)^2}$—standard deviation from the mean diameter; $d_m = \sum\beta_i d_i$—mean diameter of particles in the mixture; β_i—relative content of i fraction in sediment in the mixture, d_i—diameter of the i-th particle fraction in the mixture. These dependences can be used for river flows with smooth beds.

The impact on water flow resistance can also be associated with randomly distributed individual accumulations of coarse particles at the bottom of the river, as well as the presence of the bed forms. The latter may be a relict, and their sizes are uniquely unrelated to the flow characteristics. Therefore, it is not possible to use the available techniques to assess the contribution of the bed form component. In this case, the only possible solution is to use the empirical relation and try to link the grain roughness value with the value of the lengthwise total resistance. There were several such studies conducted by Griffiths [12], Kishi and Kuroki [18] and Engelund and Hansen [19], but only Griffiths used the materials of measurements for rivers with coarse-grained sediment. On the basis of experimental data of Jaeggi [20] and MPM, he derived the following design formula for tranquil flows:

$$\Theta_r \equiv \tau_{*r} = \tau_* - \tau_{*d} = exp(-b\tau_*{}^{-m}) \cdot (\tau_* - \tau_{*c}) \quad (7)$$

where $b = 0.142$ and $m = 0.71$ are regression coefficients respectively, Θ_r or τ_{*r} the dimensionless shear stress due to bed forms, τ_{*d} the dimensionless shear stress due to grains and τ_* the dimensionless shear stress due to both grain roughness and bed form roughness.

Given that additional resistance factors in the natural beds cannot be identified and evaluated as of today, the use of integral estimation based on the field measurements data allows for a sufficiently reliable description of the flow—channel interaction process in rivers with coarse-grained sediment.

1.3. Bed Forms Resistance—Total Lengthwise Resistance

The hydraulic resistance to water flow in rivers with fine-grained bottom sediments differs significantly from the resistance of pebble-gravel beds. The main differences are as follows.

In rivers with sandy bottoms, the share of grain roughness becomes much smaller. This is due to the relatively large values of H/d. Moreover, as shown by the results of studies, there is sediment sorting on the bed forms surface in sand beds. Therefore, the representative diameter of the bottom sediments on the pressure bed form slope is less than the value estimated for the original mixture.

The second difference is that the size of the bottom forms in the sand beds weakly depends on the change in coarseness of the bottom sediments.

Finally, due to the high mobility of the sand particles, the channel forms are transformed by the flow when water movement characteristics change. Therefore, the water flow of the river is able to regulate the roughness of its bottom in virtually the entire range of the runoff change during a calendar year.

Grishanin [14] derived the functional dependence of the formula below when exploring hydraulic resistance of quasi-uniform flows with a sand bed and developed bed form roughness (channel reaches in rectilinear reaches of the hollows):

$$C/\sqrt{g} = f(U/\sqrt[3]{gv}, B/H) \quad (8)$$

where C ($m^{1/2}/s$) is Chézy coefficient related to the roughness due to grains and bed forms, U (m/s) is mean flow velocity, g (m/s^2) acceleration of gravity, v (m^2/s) is kinematic viscosity coefficient, B (m) the channel cross-section width, and H (m) is mean riverbed hydraulic depth in river cross-section.

When these scales were chosen as the main arguments for Chézy's coefficient, it was taken into account that resistance of the flat bottom made up of the sand fraction particles is in principle affected by viscosity, while the variations of the particle diameter within the boundaries of this fraction has no detectable effect on either the resistance of the flat bottom, or on the bed form resistance.

The resulting design formula for total Chézy coefficient is written as:

$$C/\sqrt{g} = 5.25(U/\sqrt[3]{gv})^{1/2}(H/B)^{1/6} \quad (9)$$

where Chézy coefficient related to the roughness, U is mean flow velocity, g acceleration of gravity, v is kinematic viscosity coefficient, B the channel cross-section width, and H is mean riverbed hydraulic depth in river cross-section.

Formula (9) was confirmed with a high correlation coefficient for field measurement data. The second argument in the formula is the dimensionless depth of the flow or the value reciprocal of the relative channel width. From the obtained relation it follows that with equal flow velocities, hydraulic resistance to water movement is less in relatively large and deep rivers than in small and broad ones. The second conclusion is that Chézy's coefficient increases with increasing water levels in the river.

Gladkov [21], studying bed resistance at the ripples, suggested that it is associated with the kinetic energy of the flow. The main objectives of the author are based on the idea that ripple washout at low-water levels is more intense, the larger the slope of the free surface I, and in the course of washout, the average length of bed form l_r increases. This suggests a proportionality relation $l_r/H \sim I$ between the relative length of bed form l_r/H and slope I.

To test the stated objectives, the materials of field measurements and the ripples of a number of navigable rivers were used, as well as data of the "Inland Waterways Map" at the reaches of several major navigable rivers of Russia. The sample size of the field measurement data amounted to 296 measurements at the time of the study. A new dependence of Chézy's coefficient was obtained from these data:

$$C/\sqrt{g} = 18.6\left(\frac{U}{\sqrt{gH}}\right)^{3/4}\left(\sqrt[3]{\frac{gH^3}{\nu^2}}\right)^{1/8} \quad (10)$$

The formula obtained is not inferior in accuracy to design Formula (9). The field of their possible application in practice should be linked with the dimension of the hydrodynamic model used in the calculations. Formula (9) can be recommended for solving problems in a one-dimensional definition, and the design formula of Form (10) in a two-dimensional definition, respectively. The possible field of their application is limited to the set of rivers with sandy bottom sediment at flow rates ensuring the transport of sediments in the type of the bed form.

Analysis of the results shows the changing nature of the relation between Chézy's coefficient and determining factors at various morphological elements of the river channel. In reach hollows, water flow velocity represented in a dimensionless form is the main argument of water movement resistance, in quasi-uniform water movement in the rivers (9). The functional linkage with the *B/H* parameter in the resulting formula is weaker than with the flow velocity.

A different picture is observed in the ripples and shallow reach hollows at low water levels. The kinetic energy of the flow increases under these conditions, which leads to an increased linkage of Chézy's coefficient to the flow depth. Both types of channel formations—ripples and reach hollows—can be found at river bends, and both types of dependences of Chézy's coefficient can be implemented.

With the establishment of new dependences, the design practice has the description of the mechanism of interaction of the river flow and the movable bed. The resulting formulas of Chézy's coefficient implicitly consider the changing parameters of the bed forms and lengthwise energy losses caused by them with a change in the boundary water movement conditions. The use of these formulas in hydraulic calculations enables assessing the effect of water transport engineering measures on the hydraulics of the river flow and sediment transport conditions in rivers.

Further improvement of the justifying calculations in the field of sediment transport is associated with objective difficulties. Until now, the assessment of the water resistance value and sediment flux in the river channel have been studied separately. However, the value of energy losses over the length, which is estimated by the Chézy formula, enters into the water movement equation and the deformations equation within the sediment flux formula. Therefore, when analyzing the sediment transport characteristics in rivers, it is difficult to find objective and physically measurable parameters of displacement of the river alluvium particle under the action of the flowing water.

The main objective of this study is to verify different sediment transport models on the basis of previous research and to verify them based on the measurements on the alluvial stretches of rivers and a modified model of transport of **non-uniform grain-size sediments**. Implementation of the works is to allow outlining further steps of improving sediment transport models for use in the design practice. To test the stated objectives, the materials of detailed measurements of a number of navigable lowland rivers were used, as well as data of the "Inland Waterways Map" at the reaches of several major navigable rivers of Central-Eastern Europe.

2. Materials and Methods

2.1. Input Data for Testing the Model

Various sediment transport formulas were tested in the study based on the data sample of measurements carried out on rivers with different coarseness of the bottom sediments. Materials of the measurement taken on the large and medium alluvial rivers of the lowlands, in the reaches of mountain and piedmont rivers with coarse-grained bottom sediments, as well as the materials of the two series of experimental lab studies in the scaled hydraulic flume with movable bottom [22] were used for sample design. At this stage, the input data sample includes measurement results on 41 hydraulic sections of 29 different rivers of Russian territory and several former USSR republics, published in the State Inventory of Water Resources [23], as well as measurements on other Central-Eastern European rivers [17]. General characteristics of the adapted data sample are presented in Table 1.

Table 1. General characteristics of the sample of measurement materials for calibration of the sediment transport model. Adapted from: Danube and Isar [17], others [23].

River Name, Water Gauge Post and Country	Years of Measurements	Mean Water Velocity (m/s), From/To	Mean Channel Depth (m), From/To	Diameter of the Sediment Grains (mm), From/To	Water Surface Slope (‰), From/To	Number of Measurements
Danube (Pfelling, Hofkirchen—Germany)	1970, 1971 1989–1991	0.95/1.94	3.10/5.92	10.7/17.6	0.11/0.34	9
Isar (Plattling—Germany)	1988, 1989	1.44/2.12	1.48/2.61	22.5	0.85/0.89	3
Niekuchang (Russia)	1955	0.98/1.71	0.32/0.49	93.2	9.8/11	4
Burtochan (Russia)	1955	1.22/1.70	0.45/0.70	50.1	4.0/4.6	4
Chu (Tashkul Village—Kazakhstan)	1949, 1955	0.5/1.25	0.8/2.17	0.31/2.55	0.30/0.62	25
Desna (Chernigov—Ukraine)	1964, 1955	0.44/0.95	3.73/6.30	0.25/0.27	0.029/0.10	6
Moscow (Zvenigorod—Russia)	1952, 1955	0.29/1.06	0.39/3.68	1.63/1.76	0.16/0.38	16
Yenisei (Minusinsk—Russia)	1958	0.44/0.87	1.14/2.86	21.1	0.08/0.20	6
Klazma (Kovrov—Russia)	1950, 1951	0.40/1.10	3.83/6.2	0.59/2.12	0.027/0.078	13
Suda (Kurakino—Russia)	1949	0.40/0.98	1.84/2.72	3.49	0.086/0.21	10
Mologa (Ustyuzhna—Russia)	1951, 1952	0.31/0.81	1.08/3.96	1.83/2.48	0.036/0.13	13
Ob (Ogurtsovo—Russia)	1950, 1951	0.63/1.27	1.34/5.3	0.36/0.47	0.036/0.15	36
Volga (Yaroslav—Russia)	1953	0.59/0.77	6.0/7.0	0.99	0.035/0.045	3
Kara Darya (Uzbekistan)	1955–1958	1.29/2.77	0.81/1.95	39.6/56.8	2.1/4.1	25
Arys (Kazakhstan)	1957, 1955	0.46/0.87	0.39/2.99	1.72/3.66	0.28/0.63	15
Naryn (Kirghizia)	1955	1.37/2.08	2.55/3.43	47.6	0.86/1.6	12
Kurta—(Kazakhstan)	1956–1958	0.36/0.84	0.09/0.82	0.71/2.83	0.35/1.4	28
Dnieper (Kiev—Ukraine)	1959	0.39/0.78	2.96/7.0	0.34	0.041/0.065	8
Oka (Novinki—Russia)	1946	0.57/0.82	2.07/6.3	0.24	0.015/0.035	4
Charysh (Charyshsky Village—Russia)	1951	0.63/0.67	2.54/2.63	0.51	0.26/0.37	2
Togul (Togul Village—Russia)	1951	0.33/0.55	0.26/0.41	3.53	0.08/0.21	2
Burla (Habara Village—Russia)	1951	0.09	0.14	0.41	0.27	1
Ishim (Petropavlovsk—Kazakhstan)	1952	0.16/0.64	0.70/4.9	0.65	0.03/0.11	9
Chulyom (Communarca—Russia)	1952	0.88/1.07	5.4/5.7	0.7	0.09/0.092	3
Vyatka (Kirov—Russia)	1954	0.60/0.84	1.8/4.23	0.99	0.093/0.13	10
Suuremayygi (Kvissental—Estonia)	1956	0.27/0.64	3.27/4.5	4.1	0.022/0.088	9
Don (Kazan Village—Russia)	1953	0.62/1.00	5.4/7.3	0.27	0.041/0.095	7
Sheksna (Black Grow—Russia)	1949	0.23/0.56	3.97/4.97	0.37	0.003/0.022	10
Nura (Sergei Village—Kazakhstan)	1948	0.19/0.47	0.47/0.96	4.75	0.47/0.53	3
Overall outliers for the sample	-	0.09/2.77	0.09/7.3	0.25/93.2	0.003/11.0	-

All measurements published in the State Inventory of Water Resources [23] and data from the Federal Waterways Engineering and Research Institute (Bundesanstalt für Wasserbau) used by Söhngen [17] were carried out with grab samplers for collecting sediment samples from hard bottoms (sand, gravel) using a unified approved methodology by certified measuring instruments at stationary hydrological posts of the state networks. Hydraulic and morphometric characteristics within the sample vary over a rather wide range. Thus, the mean flow velocity values vary from 0.09 to 2.77 m/s; mean flow depth—from 0.09 to 7.3 m; diameter of the sediment grains—from 0.25 to 93.2 mm; free surface slopes—from 0.003 to 11.0‰, i.e., by 3660 times.

2.2. The Structure of the Transport Model for Non-Uniform Size of Grain Bottom Sediments

The design formulas of MPM [4], van Rijn [24,25], Hunziker [26,27], Parker [28,29] and Einstein [30] and their modifications (a total of 30 calculation models) were used as the main dependences for modeling the transport of **non-uniform grain-size sediments** for the purpose of their verification [31]. Ribberink models were used as known modifications of the formula of MPM [15] and the Federal Waterways Engineering and Research Institute

(BAW) [17,32]. Calculations were also performed by the van Rijn formula modified by Lagucci [33]. A number of new modifications of these formulas have been plotted based on experimental and field measurement data obtained by the author's research [21]. The calculation algorithm using the modified formula of van Rijn is presented in Figure 1. Test calculations were performed taking into account the mean diameter, and by fractions. The original formula of van Rijn [25] was derived from measurements of sediment discharge on large rivers Q_s (kg/s) and tested against 226 measurements. The range of particle diameters varied from 0.1 to 2.0 mm, depths from 1 to 20 m, and the flow velocity from 0.5 to 2.5 m/s. van Rijn's formula is written as:

$$Q_s = 0.005 \cdot Q \cdot \left(\frac{U - U_c}{\sqrt{\frac{\rho_s}{(\rho-1)gH}}} \right)^{2.4} \tag{11}$$

where Q (m^3/s) is volumetric flow rate, U *(m/s)* is mean flow velocity, U_c (m/s) the critical flow velocity, ρ_s (kg/m^3) and ρ (kg/m^3) are densities of particles and water respectively, g (m/s^2) the acceleration of gravity, v (m^2/s) is kinematic viscosity coefficient, B (m) the channel cross-section width, and H (m) is mean riverbed hydraulic depth in river cross-section.

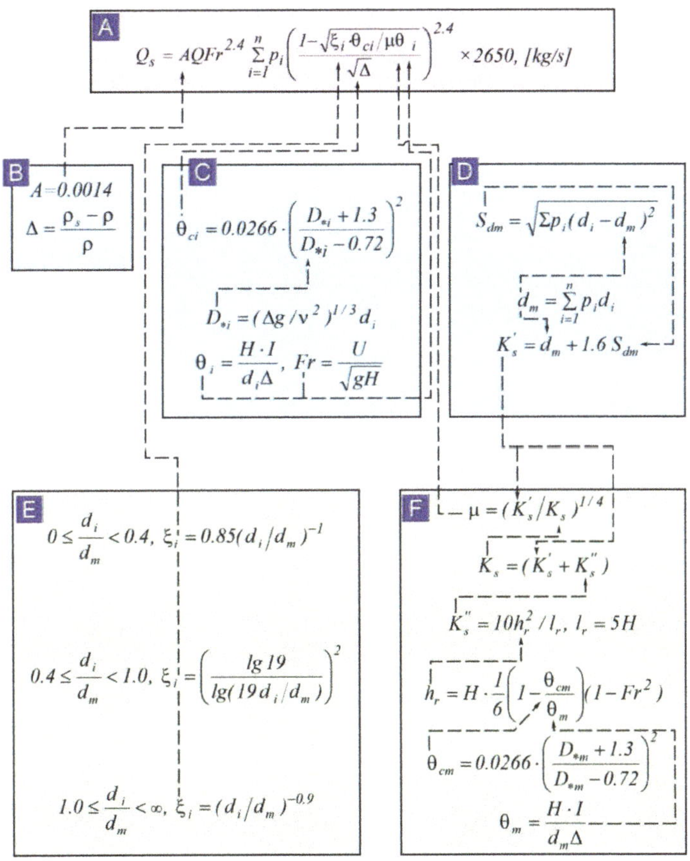

Figure 1. The modified sediment transport rate of van Rijn's formula [31]—modified—Box A, with the course of calculations of the components, where Box B is the constant term, Box C—calculation of the critical value of Shields' coefficient θ_{ci} in the modification by Knoroz [34]; Box D—calculation of

the representative diameter of bottom sediments S_{dm} as standard deviation from the mean value; Box E—calculation of the Egiazaroff [35] sheltering coefficient ξ_i with proposed conditional function by Kishi and Kuroki [18]; Box F—calculation of the ripple factor μ according to Meyer-Peter and Müller [4] in the modification by Söhngen [32]. Explanation for abbreviations: i—number of size fraction, ξ_i—coefficient, taking into account different probability of incipient motion of non-uniform sediments (according to Ribberink [15]), θ_{ci}—critical value of shear stress according to Knoroz [36], θ_i—shear stress of the i-th fraction, μ—ripple factor (according to MPM, where form roughness was determined by Yalin recommendations), ρ_s and ρ are densities of particles and water, respectively, p_i—percentage of i-th size fraction, A—constant term, D_{*i}—dimensionless particle diameter for the i-th ground particle fraction, Fr—Froude number, $K_s{}'$ and $K_s{}''$—are equivalent roughness related to grains and bed form, respectively, K_s—effective bed roughness, L_r—length of bed form, h_r—bed form height, θ_{cm}—critical value of the mobility coefficient for medium-sized grain particles, θ_m—value of the mobility coefficient for medium-sized particles, H—mean riverbed hydraulic depth in river cross-section, I—slope, d_m—mean diameter of particles in mixture.

The value of the critical velocity U_c for sand with a size of $0.5 \leq d_{50} < 2.0$ mm in van Rijn's formula is calculated according to the following relationship:

$$U_c = 8.5 d_{50}^{0.6}\, lg\left(\frac{12H}{6d_{50}}\right) \tag{12}$$

The modification of van Rijn's formula shown in Figure 1 was obtained in [31]. In this work, this formula was tested on the basis of 296 measurements on rivers (Table 1). The improvement of van Rijn's formula was achieved by solving the following issues. The developed sediment transport model uses a new expression (see Box C in Figure 1) to assess the stability conditions of sediment particles at the bottom, obtained on the basis of experimental data by Knoroz [22,34] for a wide range of grain size variation bottom sediments. The new model takes into account the effect of changing the critical conditions for the shear of particles of different-grained sediments in the mixture (see Box E on Figure 1)—the so-called Hiding-Exposure (HE) coefficient. The recommended calculation methodology is based on independent proven materials of theoretical and experimental studies carried out by Kishi and Kuroki [18], Egiazaroff [35] and Söhngen [32].

In Boxes D and F, based on the research results by Ribberink [15], an algorithm is shown for calculating the relative proportion of grain roughness, the so-called ripple factor in the form $\mu\Theta_i$. River channel parameters are calculated according to the recommendations by Yalin and Karahan [36].

During the calculations by each design formula, a regression dependence of the form $ln(Q_s)_{meas} = a + b(ln(Q_s)_{calc})$ was plotted between the measured and calculated sediment transport rates, by which the correlation of that linkage was evaluated.

The second criterion for determining the quality of calculations by each model was the requirement to keep the author's interpretation of the tested formula. The exponent at the flow velocity (mobility ratio) in the sediment discharge formula remains unchanged with the angular coefficient $b = 1$ in the regression equation.

2.3. Test Calculations of the Grain Roughness Resistance in Natural Channels

Based on the available input data sample, various formulas for estimating the grain roughness resistance in natural channels have been tested. The design formulas of Manning–Strickler, Zegzhda [7], Griffiths [8], Yalin and Karahan [36], Engelund [37] and Karim [38] were used in the calculations. Given that each of the analyzed calculation dependences has empirical coefficients the values of which depend on the research conditions in each case, the work was carried out in two stages.

At the first stage, separate formulas for calculating grain roughness of the channel were compared with each other. This approach implies certain convention, because it is not possible to distinguish the relative proportion of the grain roughness alone, and it is unclear which of the analyzed formulas should be assumed as basic for comparison. In this regard, the second stage of calculations was carried out, in which the values of tractional sediments discharge were calculated using the modified sediment discharge van Rijn formula. In this work a solution was tested, according to which the value of the mobility coefficient corresponding to the relative fraction of the granular bottom roughness was calculated, which reads as: $\Theta_d = \mu \cdot \Theta$, where μ is ripple factor, $\Theta \equiv \tau_* = \frac{\rho(U_*)^2}{(\rho_s-\rho)gd}$ is mobility coefficient or dimensionless shear stress due to grain and bed form roughness, Θ_d is mobility coefficient (dimensionless shear stress) due to grain roughness, U_*(m/s) is bed shear velocity and d (m) grain diameter. In this work, a different scientific and methodological approach was adopted. The same modified formula of van Rijn was adopted as the basic formula for testing the sediment transport model, which showed relatively better results at the previous stages of research [31]. The main difference lies in the use of a different approach to estimate the value of the ripple factor (see Boxes D and F). The relative fraction of the granular roughness of the bottom was determined using the value of the Chézy coefficient C_d calculated by the formulas of various authors (see Figures 2 and 3). The value of the mobility coefficient corresponding to the granular roughness of the bottom was found using the well-known formula:

$$\Theta_d \equiv \tau_{*d} = U_{*d}^2/\Delta g d = U^2/C_d^2 \Delta d \tag{13}$$

where $\Delta = \frac{\rho_s-\rho}{\rho}$.

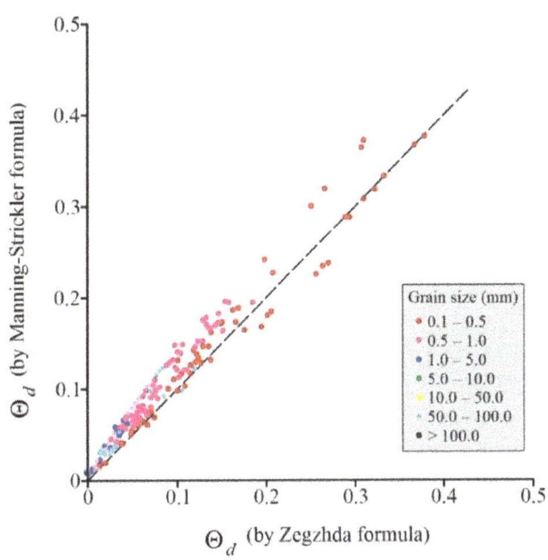

Figure 2. Comparison of the results of calculation of sediment mobility coefficient Θ_d parameter by the formulas of Manning–Strickler and Zegzhda.

Checking the quality of the model was carried out as follows. According to the modified van Rijn formula shown in Figure 1, using an array of initial data (see Table 1), the values of sediment discharge were calculated. As mentioned above, when performing these calculations, the value of the mobility coefficient Θ_d corresponding to the granular roughness of the bottom was calculated using the Chézy coefficient according to the above, Formula (13). Accordingly, Boxes D and F shown in Figure 1 were not involved in the calculations in this case. Further, the calculated values of sediment flow rates were compared with the measured sediment discharges and thus the values of empirical coefficients in the used formulas of Chézy's coefficient C_d were finalized.

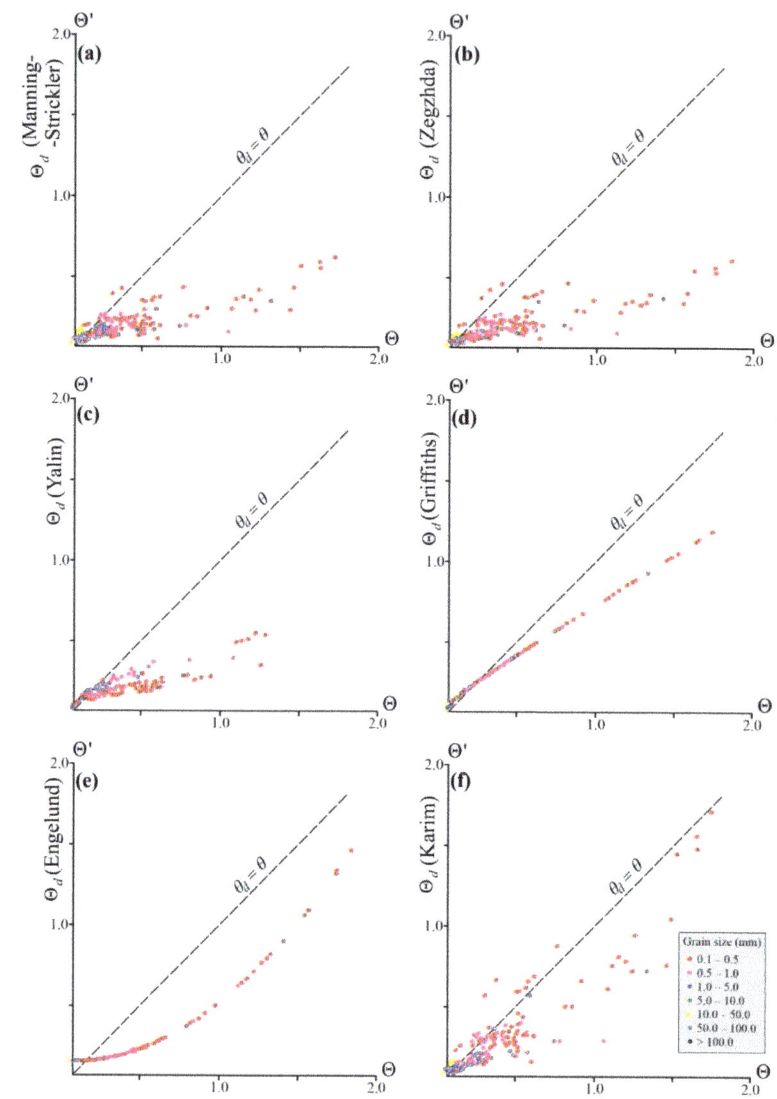

Figure 3. Results of analytical dependence of $\Theta_d = f(\Theta)$ graphs, where: (**a**)—Manning–Strickler formula; (**b**)—Zegzhda formula; (**c**)—Yalin formula; (**d**)—Griffiths' formula; (**e**)—Engelund formula; (**f**)—Karim formula.

The formulas of Manning–Strickler and Zegzhda [7] of the mobility coefficient Θ_d were compared. This allowed for certainty in the assessment of the results obtained and outlining the next steps of research in this field. On the basis of the calculations made with the available input data sample (see Table 1), the $\Theta_d = f(\Theta)$ graphs were plotted. The Θ_d value in each case was calculated from the different formulas of Chézy's coefficient for estimating the grain roughness of the bottom.

3. Results

3.1. Sediment Transport Model for Non-Uniform Size of Grain Bottom Sediments

The structure of the transport model for **non-uniform grain-size sediments** test calculations showed that of all the options considered, the best calculation results are obtained based on the original van Rijn formula [24,25], developed by the author for calculating the fine-grained sediment discharge in rivers [31]. In this work, we have made some headway. Calculations using the Manning–Strickler and Zegzhda formulas showed that when using the constant term, A = 0.001 in the calculation formula for the sediment discharge by van Rijn, the quality of calculating the sediment discharge becomes higher than that obtained at the previous stage of the study at A = 0.0014. Modification of this dependence enabled expanding the field of its possible use for a wider range of particle size of sediment of the bottom in rivers. At the same time, the calculations performed on the basis of the Strickler formula give the best results in the author's interpretation of the formula. The value of the free term in the Zegzhda formula turned out to be 1.0, which coincides with the result obtained earlier by Limerinos.

Comparison of the calculated values of the tractional sediment discharges with the measurement data indicate their satisfactory coincidence. The correlation coefficient in the regression equation was 0.887. According to the model testing results, the angular coefficient b in the regression equation was 0.997, which allowed keeping unchanged the exponent value 2.4 as set by the author in the modified formula.

The performed calculations have shown the improved modeling quality when passing from the calculation by the mean diameter to calculation by fractions. This suggests that the guidelines used to account for the ripple factor and hiding effect are correct; models for accounting for them do not conflict with the general calculation procedure and are physically grounded. The calculation results are almost independent of the input data processing technique. They can equally use the arithmetic mean and geometric mean scores of the fractional composition.

The quality of sediment discharge calculation by the MPM formula in its author's form and in its various modifications turned out to be lower than by the van Rijn formula. In all variants of the test calculations, the angular coefficient in the regression equation proved to be different from one (1.0), and was $b = 0.77$ on average. In this connection, the field of the possible use of the original dependence of MPM should be limited to the range of values of the large particle diameters, set by the authors in deriving it.

Finally, the probabilistic formula of Einstein [30] turned out to be rather satisfactory in terms of the accuracy of calculations. As for all other calculation dependences, test calculations showed relatively lower results compared with the field measurement data.

3.2. Sediment Mobility Coefficient Calculations

The factor that was not considered at this stage of research is associated with evaluating the hydraulic resistance of the channel. This is because the value of the mobility ratio (by Shields' method) was calculated on the basis of measurements of the free surface slopes. In design practice, i.e., the free surface slope value is determined by calculation at each computation step when modeling the reformations in the river channel, depending on the calculated (set) value of Chézy's coefficient. Therefore, it is not quite correct to talk about the quality of sediment transport model without solving this issue. These problems must be solved together, which was the subject of further research.

At this stage, an attempt is made to exclude the intermediate steps of computations associated with distinguishing the bed form component of energy losses over the length and to find the value of the mobility coefficient in the sediment discharge formula using Chézy's coefficient by the known formula of grain roughness.

Given that quite a number of such formulas are used in the design practice, this study undertakes to test them and develops guidelines on improving the calculation procedure for the sediment transport parameters in rivers.

The data obtained from the comparison of formulas of Manning–Strickler and Zegzhda show that the mobility coefficient calculation results by these formulas match well with one another. In the range of small values of the mobility coefficient, i.e., for relatively larger particles, the Manning–Strickler formula gives higher values in the calculations compared with the Zegzhda formula. The results of the comparison of separate formulas for calculating grain roughness of the channel are shown in Figure 2. The formulas of Manning–Strickler and Zegzhda of the mobility coefficient Θ_d were compared.

Graphs were plotted on the basis of the calculations made with the available input data sample (Table 1). The value in each case was calculated from the different formulas of Chézy's coefficient for estimating the grain roughness of the bottom. Comparison of the results of calculations is shown in Figure 3.

Analysis of the obtained results suggests the following conclusions. The closest results between themselves were received from the formulas of Manning–Strickler and Zegzhda (Figure 3a,b). The number of values that fell in the $\Theta_d \geq \Theta$ range, was smaller with the Zegzhda formula than with Strickler's formula. This indicates that the design formula of Zegzhda will give more accurate calculation results in the range of small values of the mobility coefficient than the design formula of Strickler, with other things being equal.

The results of calculations performed using the Yalin and Karahan technique [36] correlate fairly well with the data obtained with the first two formulas; however, this technique also yields conservative values of Θ_d in the range of low mobility of the particles (Figure 3c). In calculating the bed form height by the Yalin and Karahan technique, an additional parameter—a critical value of mobility coefficient Θ_{cm}—is used. In our case, its value was found from the calculation dependence based on the experimental studies of Knoroz [34].

Analytical dependence $\Theta_d = f(\Theta)$ established by Griffiths, judging on the data obtained, overstates the relative proportion of the grain roughness. This is most strongly manifested in the range of large values of the mobility factor, i.e., in relation to the small-size particles (Figure 3d). The Θ_d value calculated with this formula is significantly greater in virtually the entire range than the values obtained by the formulas of Shtrickler, Zegzhda and Yalin and Karahan. Apparently, the constant term value obtained by the author in the regression equation should be adjusted according to this calculation dependence when calculating the sediment discharge value.

Graph $\Theta_d = f(\Theta)$ based on calculations by the Engelund formula [37] has a reverse curvature in comparison with other analyzed formulas (Figure 3e). This is difficult for physical justification, and additional experimental and field measurement data should be further involved for test calculations to make a decision regarding the possible use of this formula in the calculations.

The graph obtained from calculations by the Karim formula [38] clearly shows the regions separating the mobility grade of small and relatively larger particles. This is apparently due to the fact that the author's original formula was tested by measurements on rivers with fine-grained material. However, even for these fractions, the scatter of points on the graph was rather wide (Figure 3f).

To assess the accuracy of calculations of the tractional sediment discharge using the selected calculation dependences of energy losses over the length, the second stage of calculations was carried out in the study, in which the results of calculations were compared with field measurements on the rivers using the available sample. The following method was adopted for calculations.

The modified van Rijn formula was used as a basic model for testing the sediment transport rate; it is described in this figure. The test calculations were performed for all of the above formulas of Chézy's coefficient. In the course of the calculations, the values of the parameter in the original formula were defined with all the above-described design formulas, allowing estimation of the grain roughness proportion in each case, and, if necessary, making the necessary adjustments of its value. The comparison of the calculation values of sediment transport rate with the measurements was used as an objective criterion of the performance quality of a particular design formula for estimating the grain roughness of the channel.

4. Discussion

The results obtained in the course of verification of different sediment transport models based on the field measurement data allowed outlining further steps of their improvement for use in the design practice. They are as follows in general terms. According to the carried-out analysis, it is further expedient to develop three main models based on the original formulas MPM [4], Einstein [30] and van Rijn [24,25], with an account of their possible modifications on the basis of the research by Ribberink [15], Knoroz [34] and Gladkov [39,40]. In such a combination, these models give better results when verified by field observations on the rivers in a wide variation range of hydraulic and morphometric parameters.

Despite certain progress in recent years was achieved through research into the energy loss valuation problem for natural channels with a movable bottom, there is still a number of unresolved issues. The main difficulty is that the stream flow interacting with a movable bottom independently regulates the boundary conditions of its movement through the degrees of freedom it has. In this case, the nature and direction of changes in the flow channel system are still not clearly understood [41].

With reference to the flows with deformable channels, the distribution of hydraulic resistance along the length depending on contributing factors such as grain roughness of the bottom, bed form resistance and channel form resistance, is quite difficult, as the contribution of each of the components varies both over the river length and depends on time, more specifically on the hydrological cycle. Therefore, the need for distinguishing the relative proportion of any type of water movement resistance is always a certain convention, and in each case depends on the problems to be solved. At the reaches of the rivers with movable sediments in the bottom, it is more preferable to use the free surface elevations of the new dependences of the lengthwise hydraulic resistance in the calculations, obtained based on the established relation between Chézy's coefficient and water flow rate [14,21]. The structure of these formulas allows considering the feedback mechanism in the river flow—movable channel system and, on this basis, a more reliable estimate of the river flow response to natural and artificial changes in the river channel.

Calculations made by the formulas of Manning–Strickler and Zegzhda [7] have shown that the quality of calculating the sediment discharge is improved when using the constant term A = 0.001 in the sediment discharge design formula of van Rijn, than was obtained during the previous stage of the study with A = 0.0014 [31]. The calculations based on the Strickler formula [6] give the best results in the author's interpretation of the formula. The constant term value in Zegzhda formula [7] was found to be 1.0, in agreement with results previously obtained by Limerinos [13].

Calculations with all other design formulas were performed for these two cases, i.e., with A = 0.0014 and A = 0.001, and only the proportionality coefficients in the relevant lengthwise energy loss formulas were subject to adjustment in assessing the proportion of the grain roughness of the channel. According to the calculation results, when using the calculation dependences of Griffiths [8], Engelund [37] and Karim [38] for calculating the tractional sediment discharge, the author's empirical coefficients in these formulas need adjustment to improve the quality of calculations. Based on the tests performed in the evaluation of the relative share of the grain roughness with these formulas, it is necessary to investigate further the possibility of their use in the design practice, taking into account the need for adjustment of the authors' coefficient values in the equations of regression analysis. This would allow for further improvement of the sediment transport model to enhance the quality and reliability of the channel forecasts. However, at this stage of our research, it can be concluded that these results can be successfully used during the configuration of numerical models.

Egiazarov [35], in developing the logarithmic rule of the distribution of flow velocities in the bottom layer, established a dependence between the relative mobility coefficient for the i-th soil fraction—normalized by its average Θ_{cm} value—and the corresponding relative value of the $\frac{d_i}{d_m}$ particle diameter. This dependence has received extensive experimental verification.

More recent studies by Wilcock and Crowe [42] and Bakke et al. [43] corroborate the well-known fact that the mobility of smaller particles in a mixture decreases due to their "shading" by larger particles, and vice versa. The probability of large particles shifting in a mixture is higher than in the case of homogeneous particles of the same diameter. The results of the studies carried out by Gladkov and Söhngen [31] showed that when moving from the average diameter calculation to fractional calculation, the modelling quality increases. This indicates that the recommendations to take into account the ripple factor and the "hiding" effect are correct, and the models which take them into account are consistent with the general calculation method and are physically justified. At the same time, the outcome of the calculation practically does not depend on how the input data is processed—both the arithmetic mean and geometric mean estimates of the fractional composition of bottom sediments can be used on a par [31].

The quality of the results obtained in creating and verifying numerical models of sediment transport discharge in rivers depends to a large extent on the reliability of the field data used. Traditional methods of measuring sediment discharge with bottom grab samplers show a rather large error, due to their high randomness. The results obtained in the work by Gladkov and Söhngen [31] and in the present work are based on measurement data from bottom grab samplers. However, it is necessary to develop other measurement methods. This can be accomplished by research based on the use of the methodology for calculating the sediment transport discharge using the parameters of bottom ridges (small bed forms), which can objectively lead to an increase in the quality of forecasts regarding river channel transformations on [44–46].

5. Conclusions

The main recommendations for using the results obtained in hydraulic calculations and modeling of channel reformations in rivers are as follows.

When performing hydraulic calculations to determine free surface elevations along the length of a river with coarse-grained bottom sediments, it is advisable to use the calculation dependences of the type of Manning–Strickler and Zegzhda. With the availability of the measured free surface slopes, a series of test calculations is required to specify the constant term value in the initial dependence or the value of the effective height of the roughness protrusions on the bottom of the river flow, which will allow accounting for additional water movement resistance factors.

In the sediment transport simulation and calculation of channel deformations in the rivers, it is expedient to use the calculation dependences of Chézy's coefficient obtained for evaluating the grain roughness of bottom sediments, or dependences of the form $\Theta_d = f(\Theta)$ based on the measurement research materials. In the latter case, it is necessary to have information about the measured slopes of the free surface over the river length, which permits distinguishing the relative proportion of the grain roughness of the bottom from the full value of the lengthwise energy loss.

The performed calculations have shown that the modified van Rijn formula gives the best results among all the considered variants of test calculations, with the availability of the measured free surface slopes at the studied reach of the river with movable sediments in the bottom in comparison with the data of tractional sediment transport rate measurements. The constant term in this formula is A = 0.0014 when using the Einstein–Yalin technique for estimating the bed form component of Chézy's coefficient and the method of Ashid–Egiazarov–Zengen for the fractional calculation of **non-uniform grain-size sediments** discharge.

The results of calculations performed on the basis of design formulas for estimating the grain roughness of the bottom by using the formulas of Manning–Strickler and Zegzhda show that the application of these formulas to determine the mobility coefficient Θ_d in the tractional sediment discharge formula may be advised if free surface slope measurements are not available. When using the van Rijn formula as a sediment transport model in this case, the constant term value was found to be A = 0.001. The Manning–Strickler formula can be used in the author's interpretation, and the constant term in the Zegzhda formula should be taken as equal to 1.0.

The group of calculation dependences based on the use of functional linkage of the form $\Theta_d = f(\Theta)$ established by Griffiths, Engelund and Karim, can improve the quality of sediment transport modeling. The accuracy of calculations by these formulas is roughly the same but remains lower than the Einstein–Yalin formulas.

The problem that has been investigated at the present stage is closely related to all of the listed issues and is devoted to the assessment of the granular roughness of the bottom. The results obtained in the course of the research made it possible to obtain new data, with the help of which it seems possible to improve the quality and reliability of modeling the sediment transport in natural channel flows.

Author Contributions: Conceptualization, G.G. and M.H.; methodology, G.G., M.H., Z.B. and P.B.; validation, G.G. and Z.B.; investigation, G.G. and P.B.; writing—original draft preparation, G.G. and M.H.; writing—review and editing, M.H.; visualization, G.G. and P.B. All authors have read and agreed to the published version of the manuscript.

Funding: The APC was funded by Kazimierz Wielki University in Bydgoszcz and Admiral Makarov State University of Maritime and Inland Shipping, St. Petersburg out under a bilateral agreement concluded in 2019.

Institutional Review Board Statement: Not applicable.

Informed Consent Statement: Not applicable.

Data Availability Statement: The data presented in this study are available on the request from the corresponding author.

Acknowledgments: Thanks to the Admiral Makarov State University of Maritime and Inland Shipping for providing a hydraulic laboratory with flumes for experiments during Michal Habel's research stay. We would like only to thank the four anonymous reviewers for constructive comments that substantially improved our paper.

Conflicts of Interest: The authors declare no conflict of interest.

References

1. Leopold, L.B.; Maddock, T. *The Hydraulic Geometry of Stream Channels and Some Physiographic Implications*; United States Government Printing Office: Washington, DC, USA, 1953; p. 57.
2. Grishanin, K.V. *Dinamika Ruslovykh Potokov*; Gidrometeoizdat: Leningrad, Russia, 1979. (In Russian)
3. Manning, R. On the flow of water in open channels and pipes. *Trans. Inst. Civ. Eng. Irel.* **1891**, *20*, 161–207.
4. Meyer-Peter, E.; Müller, R. Formulas for bed-load transport. In Proceedings of the 2nd Meeting of the International Association for Hydraulic Structures Research, Delft, The Netherlands, 7 June 1948; pp. 39–64.
5. Grishanin, K.V. Hydraulic resistance of sand beds. In Proceedings of the Second International Symposium on River Sedimentation, Nanjing, China, 11–16 October 1983; pp. 234–238.
6. Strickler, A. Beiträge zur Frage der Geschwindigkeitsformel und der Rauchigkeitszahlen für Ströme, Kanäle und geschlossene Leitungen. *Mitt. Eidgenöss. Amtes Wasserwirtsch.* **1923**, *16*, 3–77.
7. Zegzhda, A.P. *Gidravlicheskie Poteri na Trenie v Kanalakh i Truboprovodakh*; Gosstroiizdat: Leningrad, Russia, 1957.
8. Griffiths, G.A. Flow resistance in coarse gravel bed rivers. *J. Hydraul. Div.* **1981**, *107*, 899–918. [CrossRef]
9. Bathurst, J.C. Flow resistance estimation in mountain rivers. *J. Hydraul. Eng.* **1985**, *111*, 4. [CrossRef]
10. Bray, D.I. Estimating average velocity in gravel- bed rivers. *J. Hydraul. Div.* **1979**, *105*, 9.
11. Graf, W.H. *Flow Resistance for Steep, Mobile Channels*; Comm. Lab. d'Hydraul. EPEL: Lausanne, Switzerland, 1987; Volume 54, pp. 1–12.
12. Griffiths, G.A. Form resistance in gravel channels with mobile beds. *J. Hydraul. Eng.* **1989**, *115*, 340–355. [CrossRef]

13. Limerinos, J.T. Determination of Manning coefficient from measured bed roughness in natural channels. *Geol. Surv. Water Supply Paper* **1970**, 1898-B. [CrossRef]
14. Grishanin, K.V. *Gidravlicheskoe Soprotivlenie Estestvennykh Rusel*; Gidrometeoizdat: Leningrad, Russia, 1992.
15. Ribberink, J.S. *Mathematical Modelling of One-Dimensional Morphological Changes in Rivers with Non-Uniform Sediment*; Delft University of Technology: Delft, The Netherlands, 1987.
16. Day, T.J. *A Study for the Transport of Graded Sediments*; Hydraulics Research Station: Wallingford, UK, 1980; p. 10.
17. Söhngen, B.; Kellermann, J.; Loy, G.; Belleudy, P. Modelling of the Danube and Isar Rivers morphological evolution. Part I: Measurements and formulation. In Proceedings of the 5th International Symposium on River Sedimentation, Karlsruhe, Germany, 7–12 April 1992; Volume 3, pp. 1175–1207.
18. Kishi, T.; Kuroki, M. Bed Froms and Resistance to Flow in Erodible-Bed Channels—Hydraulic Relations for Flow over Sand Waves (in Japanese). *Bull. Fac. Eng. Hokkaido Univ.* **1973**, *67*, 1–23.
19. Engelund, F.; Hansen, E. *A Monograph on Sediment Transport in Alluvial Streams*; Technisk Forlag: Copenhagen, Denmark, 1967.
20. Jaeggi, M.N.R. Formation and Effects of Alternate Bars. *J. Hydraul. Eng.* **1984**, *110*, 142–156. [CrossRef]
21. Gladkov, G.L. Hydraulic resistance in natural channels with movable bed. In Proceedings of the Int. Symp. East-West North-South Enc. State Art River Eng. Methods Des. Philos., St. Petersburg, Russia, 11–15 May 1994; Volume 1, pp. 81–91.
22. Gladkov, G.L. Obespecheniye Ustoychivosti Rusel Sudokhodnykh Rek Pri Dnouglublenii i Razrabotke Ruslovykh Karyerov. Master's Thesis, Admiral Makarov State University of Maritime and Inland Shipping, St. Petersburg, Russia, 1996.
23. *Ezhegodnye Dannye o Rezhime i Resursakh Poverkhnostnykh Vod Sushi (State Water Cadaster. Annual Data on the Regime and Resources of Continental Surface Waters)*; T.8. Vyp.0-7, 0-4.9; Gidrometeoizdat: Leningrad, Russia, 1948–1958.
24. Van Rijn, L.C. Sediment transport, Part 1: Bed load transport. *J. Hydraul. Eng.* **1984**, *110*, 1431–1456. [CrossRef]
25. Van Rijn, L.C. Sediment transport, part III: Bed forms and alluvial roughness. *J. Hydraul. Eng.* **1984**, *110*, 1733–1754. [CrossRef]
26. Hunziker, R. Natürliche Deckschichtbildung in Fliessgewässern, Modellansätze für fraktionellen Geschiebetransport. In *OEWAV-IWI- Seminar—Schriften-reihe des Österreichischen Wasser-und Abfall-Wirtschaftsverbandes*; ÖWAV: Innsbruck, Austria, 1996; Volume 105, pp. 43–62.
27. Hunziker, R. *Fraktionsweiser Geschiebetransport. Mitteilung der Versuchsanstalt für Wasserbau, Hydrologie und Glaziologie*; ETH: Zürich, Switzerland, 1995; Volume 138, p. 209.
28. Parker, G.; Klingeman, P.C.; McLean, D.G. Bedload and size distribution in paved gravel-bed streams. *J. Hydaul. Div. Proc. ASCE* **1982**, *108*, 544–571.
29. Parker, G.; Sutherland, A.J. Fluvial armor. *J. Hydraul. Res.* **1990**, *28*, 529–544. [CrossRef]
30. Einstein, H.A. The bed-load function for sediment transportation in open channel flows. In *Tech. Bull. No. 1026*; U.S. Department of Agriculture: Washington, DC, USA, 1950; p. 71.
31. Gladkov, G.L.; Söhngen, B. Modellirung des Geschiebetransports mit unterschiedlicher Korngröße in Flüssen. In *Mitteilungsblatt der Bundesanstalt für Wasserbau*; Nr. 82. Bundesanstalt für Wasserbau: Karlsruhe, Germany, 2000; pp. 123–130.
32. Söhngen, B.; Kellermann, J. *1D-Morphodynamische Modellierung großer Flußstrecken*; Wasserbauliches Kolloquium: Darmstadt, Germany, 1996; p. 36.
33. Laguzzi, M. *Modelling of Sediment Mixtures*; Waterloopkundig Laboratorium: Delft, The Netherlands, 1994.
34. Knoroz, V.S. Nerazmyvayushchaya skorost' dlya nesvyaznykh gruntov i faktory, ee opredelyayushchie. *Izv. VNIIG* **1958**, *59*, 62–81.
35. Egiazaroff, I.V. Calculation of nonuniform sediment concentrations. *Proc. ASCE Hydraul. Div.* **1965**, *91*. [CrossRef]
36. Yalin, M.S.; Karahan, E. Steepness of sedimentary dunes. *J. Hydraul. Div.* **1979**, *105*, 381. [CrossRef]
37. Engelund, F. Hydraulic resistance of alluvial streams. Closure of discussions. *Proc. ASCE Hydraul. Div.* **1967**, *93*, 287–296.
38. Karim, F. Bed configuration and hydraulic resistance in alluvial-channel flows. *J. Hydraul. Eng.* **1995**, *121*, 15–25. [CrossRef]
39. Gladkov, G.L.; Zhuravlev, M.V. Hydraulic resistance to water flow and sediment transport in rivers. In *Vestnik Gosudarstvennogo Universiteta Morskogo i Rechnogo Flota Imeni Admirala S. O. Makarova, 11.6*; Admiral Makarov State University of Maritime and Inland Shipping: St. Petersburg, Russia, 2019; pp. 1044–1055. [CrossRef]
40. Gladkov, G.L. Studying the granular roughness of river channels bottom. In *Vestnik Gosudarstvennogo Universiteta Morskogo i Rechnogo Flota Imeni Admirala S.O. Makarova, 12.2*; Admiral Makarov State University of Maritime and Inland Shipping: St. Petersburg, Russia, 2020; pp. 336–346. [CrossRef]
41. Gladkov, G.L.; Chalov, R.S.; Berkovich, K.M. Gidromorfologiya Rusel Sudokhodnykh Rek: Monografiya. Available online: https://e.lanbook.com/book/116365 (accessed on 20 February 2021).
42. Wilcock, P.R.; Crowe, J.C. Surface-based transport model for mixed-size sediment. *J. Hydraul. Eng.* **2003**, *129*, 120–128. [CrossRef]
43. Bakke, P.D.; Sklar, L.S.; Dawdy, D.R.; Wang, W.C. The Design of a Site-Calibrated Parker–Klingeman Gravel Transport Model. *Water* **2017**, *9*, 441. [CrossRef]
44. Kopaliani, Z.D.; Petrovskaya, O.A. Database "Measurement Data of the Hydraulic Characteristics of Sediment Transport in Large, Small, and Medium Plain Rivers (Certificate of State Registration of the Database no 2017620992), Official Bulletin of the "Computer Software Database Integrated Circuit Topologies" (ISSN 2313-7487). Available online: http://www1.fips.ru/wps/PA_FipsPub/res/BULLETIN/PrEVM/2017/09/20/INDEX.HTM (accessed on 20 February 2021).

45. Kopaliani, Z.D.; Petrovskaya, O.A. Baza dannykh "Dannyye izmereniy gidravlicheskikh kharakteristik transporta donnykh nanosov v gidravlicheskikh modelyakh gornykh rek i lotkovykh eksperimentakh" (Database "Measurement data of the sediment transport hydraulic characteristics in hydraulic models of mountain rivers and flume experiments"), Certificate of State Registration of the Database no. 2017620878), Official Bulletin of the "Computer Software Database. Integrated Circuit Topologies" (ISSN 2313-7487). Available online: http://www1.fips.ru/wps/PA_FipsPub/res/BULLETIN/PrEVM/2017/08/20/INDEX.HTM (accessed on 20 February 2021).
46. Belikov, V.V.; Borisova, N.M.; Fedorova, T.A.; Petrovskaya, O.A.; Katolikov, V.M. On the Effect of the Froude Number and Hydromorphometric Parameters on Sediment Transport in Rivers. *Water Resour.* **2019**, *46*, S20–S28. [CrossRef]

water

MDPI

Article

Improving the 2D Numerical Simulations on Local Scour Hole around Spur Dikes

Chung-Ta Liao [1,*], Keh-Chia Yeh [1], Yin-Chi Lan [1], Ren-Kai Jhong [1] and Yafei Jia [2]

[1] Disaster Prevention & Water Environment Research Center, National Yang Ming Chiao Tung University, 1001 University Road, Hsinchu City 30010, Taiwan; kcyeh@mail.nctu.edu.tw (K.-C.Y.); lanyinchi@gmail.com (Y.-C.L.); tkuxdwang@gmail.com (R.-K.J.)

[2] National Center for Computational Hydroscience and Engineering, University of Mississippi, 2301 S. Lamar Blvd, Oxford, MS 38655, USA; jia@ncche.olemiss.edu

* Correspondence: zeromic@gmail.com; Tel.: +886-2-571-2121

Abstract: Local scour is a common threat to structures such as bridge piers, abutments, and dikes that are constructed on natural rivers. To reduce the risk of foundation failure, the understanding of local scour phenomenon around hydraulic structures is important. The well-predicted scour depth can be used as a reference for structural foundation design and river management. Numerical simulation is relatively efficient at studying these issues. Currently, two-dimensional (2D) mobile-bed models are widely used for river engineering. However, a common 2D model is inadequate for solving the three-dimensional (3D) flow field and local scour phenomenon because of the depth-averaged hypothesis. This causes the predicted scour depth to often be underestimated. In this study, a repose angle formula and bed geometry adjustment mechanism are integrated into a 2D mobile-bed model to improve the numerical simulation of local scour holes around structures. Comparison of the calculated and measured bed variation data reveals that a numerical model involving the improvement technique can predict the geometry of a local scour hole around spur dikes with reasonable accuracy and reliability.

Keywords: numerical simulation; local scour; spur dike; angle of repose

Citation: Liao, C.-T.; Yeh, K.-C.; Lan, Y.-C.; Jhong, R.-K.; Jia, Y. Improving the 2D Numerical Simulations on Local Scour Hole around Spur Dikes. *Water* **2021**, *13*, 1462. https://doi.org/10.3390/w13111462

Academic Editor: Vlassios Hrissanthou

Received: 21 January 2021
Accepted: 21 May 2021
Published: 23 May 2021

1. Introduction

Steep slopes and severe bed changes during floods are the common characteristics of the rivers in Taiwan. Floods often cause channel incision, bank erosion, and meandering migration and endanger the safety of structures or river protection works. Spur dikes are protections commonly used for river engineering to lead water flow and reduce bank erosion. However, the local scour around spur dikes influences the stability of the dikes' foundation. To reduce the risk of foundation failure, the understanding of local scour phenomenon around hydraulic structures in rivers is important.

One can understand the formation mechanism of scour hole by the flow field and sediment movement behavior of a single cylinder pier. Figure 1 displays the scouring process around a pier structure to the front of the scour hole. Generally, the sediment entrainment and transport around the scour hole were most related to the turbulence in the approach flow. Through strong vertical flows, the turbulence fluctuations in the water are brought into the sediment layer. The turbulence fluctuation near the sediment layer enhanced the ability of the water to entrain sediment. When scour hole occurs gradually around the hydraulic structure, turbulence fluctuation in the approach flow contributes to the scouring process where there is sediment particles' pick up, trajectory, deposition, and bed evolution processes with a mixed non-equilibrium sediment transport. Moreover, the characteristics of sediment, bed load with the gravity projection, and the bed geometry of a slope have crucial effects on incipient sediment motion. The sliding effect of scour hole occurs when a local bed slope is higher than the angle of repose. The angle of repose is the steepest angle of descent relative to the horizontal bed to which the bed material can be piled without slumping. At this angle, the bed material on the slope face is on the verge of sliding.

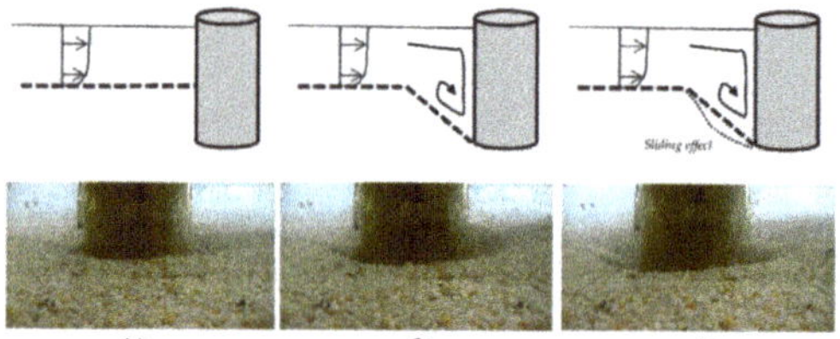

Figure 1. Scouring process around a pier structure. A laboratory experiment was conducted in the Hydraulics Laboratory of the Department of Civil Engineering, Dokuz Eylul University, Turkey [1]. (**a**) Initial state before the scour occurred. (**b**) Local scour hole that occurred around the pier. (**c**) Occurrence of the sliding effect in the front of the scour hole when the local bed slope was higher than the angle of repose.

Most mobile-bed models proposed in the past few decades have focused on the sediment transport of an alluvial channel. Jia et al. [2,3] developed a two-dimensional (2D) mobile-bed model, known as CCHE2D numerical model, which is a numerical system for modeling 2D non-steady turbulent river flow, sediment transport, and water quality prediction. The model was also designed for use in multiprocess simulation of surface water with complex geometry, such as riverbed geometry; of bank erosion with both uniform and non-uniform sediment; and of meander river migration. The model was validated through many physical experiments and field cases in the development process.

The CCHE2D numerical model can be used for river engineering and design of hydraulic structures, such as grade control structures and dikes. In the review of past research on mobile-bed models, Lai et al. [4] proposed a 2D model, known as SRH-2D numerical model that was employed to simulate the bed evolution that occurred downstream of the Chi-Chi weir in the Choushui River, Taiwan. Engineering plans pertaining to scour prevention and treatment were proposed and simulated using the model. Liao et al. [5] developed a 2D mobile-bed model, known as EFA2D numerical model that considers the non-equilibrium suspended sediment concentration profile with a bedrock erosion mechanism for natural rivers.

In the review of the research on the mechanism of modifying sediment transport behavior, Juez et al. [6–8] discuss the numerical assessment of bedload formulations for transient flow. The dam break flows over dry/wet initial conditions and erodible channel of experimental cases are simulated by considering the improved model, called Smart CFBS (Smart Combined Friction and Bed Slope model). The study has allowed a detailed analysis of the relative behavior to be performed in 2D situations of different sediment discharge formulas that were derived from 1D laboratory cases, and use of the Smart CFBS formula is suggested, regardless of the hydro-morphodynamic situation. The bedload transport over steep slopes with gravity projection has been studied and embedded in a non-trivial way, not only in the sediment transport rate but also in hydrodynamic equations. It is noted that the gravity effects play an important role on the sediment transport magnitude.

Horvat et al. [9] developed a 2D model that couples the active layer and multiple size-class approaches for modeling sediment transport by using an enhanced advection algorithm. The focus of their study was improvement of advection computation from the point of view of numerical methods by reducing the limitation of simulation time. The study proposed the advection scheme that allows for larger time steps in simulation. This restriction was overcome by introducing modifications to the characteristic method. The model was assessed and validated using field measurements conducted on the Danube River.

The 3D mobile-bed model has the capability to simulate the local scouring process. Jia and Wang [10] simulated 3D flow in a preformed scour hole. A detailed comparison of their simulation results and measurements were presented. Nagata et al. [11] developed a 3D model to simulate flow and bed deformation around river hydraulic structures. The model solved the fully 3D, Reynolds-averaged Navier–Stokes equation that was expressed in a moving boundary-fitted coordinate system to calculate the flow field of water and bed surfaces at various times. A non-linear k–ε turbulence model was employed to predict the flow near the structure, where 3D flow was dominant. Burkow and Griebel [12] conducted a numerical simulation of a fully 3D fluid flow to reproduce the current-driven sediment transport processes. The scouring at a rectangular obstacle was investigated, and the typical sedimentary processes and the sedimentary form of a scour mark were well captured by the simulation.

Castillo and Carrillo [13] studied the scour produced by spillways and outlets of a dam due to the operation of free surface spillways and half-height outlets by using three complementary procedures: obtaining empirical formulas using models and prototypes, using a semiempirical methodology based on the pressure fluctuation–erodibility index, and employing computational fluid dynamics simulations. Jia et al. [14] proposed a local scour scheme for a 3D mobile-bed model and successfully applied the model to the case of a bridge pier under non-uniform sediment conditions. The sediment entrainment, transport, and turbulence around a bridge pier are simulated. The turbulence fluctuations in the water are brought into the sediment layer through strong vertical flows. The ability of the flow to entrain sediment is enhanced by the intruding turbulence fluctuation near the sediment layer. This enhancement could be considered in 3D model by using additional shear stress related to the intruding turbulence fluctuation. However, most 3D models are still limited by the computational efficiency of the model in filed cases.

The sediment entrainment and transport around the scour hole are related to the turbulence fluctuation, characteristics of sediment, bed-load with the gravity projection, and the bed geometry of a slope that have crucial effects on incipient sediment motion. Yang et al. [15] investigated the angle of repose of non-uniform sediment by considering the concept of exposure degree to account for the hidden and exposed mechanism of non-uniform sediment transport. Meng et al. [16] proposed the angle of repose for uniform and non-uniform sediment. The angle of repose is based on the compacted friction relationship between mud and sand particles. Sediment dynamics holds that, when water flows over sediment particles on a loose surface, intersediment collision resists the fluid shear stress. This is consistent with underwater flow and sediment dynamics. Al-Hashemi and Al-Amoudi [17] reviewed the angle of repose of granular materials. This angle is an essential parameter for understanding the microbehavior of granular material and then relating this behavior to the material's macrobehavior.

In general, both an empirical formula and a numerical model can be used to evaluate the equilibrium scour depth around a hydraulic structure. Most of the empirical formulas have limitations to calculate the scouring depth or hole shape changing over time. This phenomenon and mechanism must be analyzed using the 2D or 3D numerical models. Although the empirical formula is relatively convenient to use, the numerical model is often complicated to use in terms of practical applications, especially when it is a 3D mobile-bed model. Therefore, obtaining a balance between these two approaches is crucial.

In this study, an empirical formula for calculating the sediment repose angle [16] and the bed geometry adjustment mechanism are integrated into the CCHE2D numerical model to improve the simulation of the local scour hole around a hydraulic structure. The model is calibrated and validated by simulating an experimental case [18] involving multiple spur dikes. Data are collected on flow depth, bed material grain size, scour depth around structures, and scour hole shape. These basic data are used to calibrate and validate the numerical model. By improving the 2D numerical simulations on local scour hole around spur dike, it will provide practical application value in terms of model efficiency and speed.

2. Numerical Model

The CCHE2D numerical model is a general surface water flow model. It simulates dynamic processes of flow and sediment transport in natural rivers. The CCHE2D numerical model has been developed at the National Center for Computational Hydroscience and Engineering (NCCHE), the University of Mississippi, for more than 20 years, and the developers continue to improve the model in response to different real problems.

The CCHE2D numerical model comprises flow and sediment transport modules. The governing equations of the flow module are 2D depth-integrated Reynolds equations. The free surface elevation of the flow is calculated using the depth-integrated continuity equation. Two methods for calculating the eddy viscosity exist in the model: the depth-integrated parabolic eddy viscosity formula and depth-integrated mixing length eddy viscosity model [19,20]. The sediment transport module comprises the suspended sediment transport and bed load transport simulations for non-uniform sediment. More details of the flow and sediment transport modules can be obtained in the CCHE2D technical report and verification and validation test documentation [2,3].

2.1. Flow Module

The governing equations of the flow are 2D depth-integrated Reynolds equations expressed in the Cartesian coordinate system and are as follows:

$$\frac{\partial u}{\partial t}+u\frac{\partial u}{\partial x}+v\frac{\partial u}{\partial y}=-g\frac{\partial \eta}{\partial x}+\frac{1}{h}\left(\frac{\partial h\tau_{xx}}{\partial x}+\frac{\partial h\tau_{xy}}{\partial y}\right)-\frac{\tau_{bx}}{\rho h}+f_{Cor}v \quad (1)$$

$$\frac{\partial v}{\partial t}+u\frac{\partial v}{\partial x}+v\frac{\partial v}{\partial y}=-g\frac{\partial \eta}{\partial y}+\frac{1}{h}\left(\frac{\partial h\tau_{yx}}{\partial x}+\frac{\partial h\tau_{yy}}{\partial y}\right)-\frac{\tau_{by}}{\rho h}-f_{Cor}u \quad (2)$$

where u and v are the depth-integrated velocity components in the x and y directions, respectively; t is the time; g is the gravitational acceleration; η is the water surface elevation; ρ is the density of water; h is the local water depth; f_{Cor} is the Coriolis parameter; τ_{xx}, τ_{xy}, τ_{yx}, and τ_{yy} are the depth-integrated Reynolds stresses, shown as Figure 2; and τ_{bx} and τ_{by} are the shear stresses on the bed and flow interface.

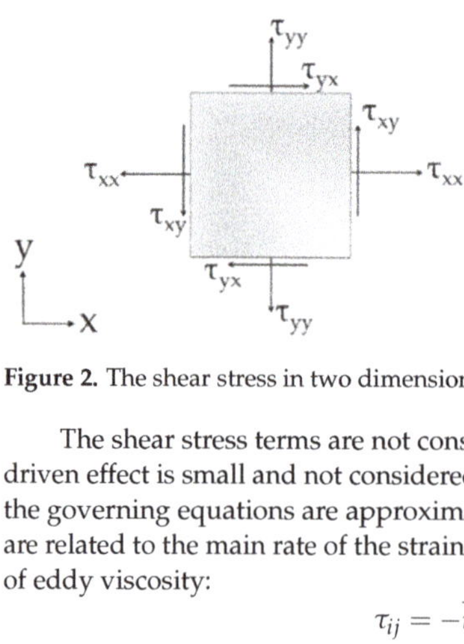

Figure 2. The shear stress in two dimensions.

The shear stress terms are not considered at the water surface because the wind shear driven effect is small and not considered in the model. The turbulence Reynolds stresses in the governing equations are approximated according to the Bousinesq's assumption that are related to the main rate of the strains of the depth-averaged flow field with a coefficient of eddy viscosity:

$$\tau_{ij}=-\overline{u_i'u_j'}=\nu_t(u_{i,j}+u_{j,i}) \quad (3)$$

Free surface elevation of the flow was calculated using the depth-integrated continuity equation:

$$\frac{\partial h}{\partial t}+\frac{\partial uh}{\partial x}+\frac{\partial vh}{\partial y}=0 \quad (4)$$

Two methods for calculating eddy viscosity exist in the model: the depth-integrated parabolic eddy viscosity formula and depth-integrated mixing length eddy viscosity model. First, the eddy viscosity coefficient ν_t is calculated using the depth integrated parabolic eddy viscosity formula:

$$\nu_t = A_{xy} C_s \kappa u_* h \tag{5}$$

$$C_s = \int_0^1 \varsigma(1-\varsigma)d\varsigma = \frac{1}{6} \tag{6}$$

where C_s is the integration constant, u_* is shear velocity, κ is the von Karman's constant (0.41) and ς is the relative depth of the flow. A_{xy} is a coefficient to adjust the value of the eddy viscosity. Its default value is set to 1 and it can be adjusted by users from 1–10. In addition to this approach, a depth integrated mixing length eddy viscosity model is also available:

$$\nu_t = \bar{l}^2 \sqrt{2\left(\frac{\partial u}{\partial x}\right)^2 + 2\left(\frac{\partial v}{\partial y}\right)^2 + \left(\frac{\partial u}{\partial y} + \frac{\partial v}{\partial x}\right)^2 + \left(\frac{\overline{\partial U}}{\partial z}\right)^2} \tag{7}$$

$$\bar{l} = \frac{1}{h}\int \kappa z \sqrt{\left(1-\frac{z}{h}\right)}dz = \kappa h \int_0^1 \varsigma\sqrt{(1-\varsigma)}d\varsigma \approx 0.267\kappa h \tag{8}$$

where the depth integrated velocity gradient along vertical coordinate $\overline{\partial U}/\partial z$ is introduced to account for the effect of turbulence generated from the bed surface. The eddy viscosity defined by Equation (7) would be zero in the uniform flow condition without this term. It is determined in the way that eddy viscosity shall be the same as that of the uniform flow in the absence of other terms. Assuming that the flow is of logarithmic profile along the depth of the water, the total velocity U of the vertical gradient should be:

$$\frac{\overline{\partial U}}{\partial z} = \frac{u_*}{\kappa z} = \frac{1}{h}\int \frac{\partial U}{\partial z} dz \tag{9}$$

2.2. Sediment Transport Module

The sediment transport module performs suspended sediment transport and bed load transport simulations for non-uniform sediment. The module is also designed to simulate a channel bed with large slopes and curved channel secondary flow effects. The capability of simulating the non-equilibrium sediment transport is achieved by solving the following equations.

Suspended load:

$$\frac{\partial(hC_k)}{\partial t} + \frac{\partial(uhC_k)}{\partial x} + \frac{\partial(vhC_k)}{\partial y} = \frac{\partial}{\partial x}\left(\varepsilon_s h \frac{\partial C_k}{\partial x}\right) + \frac{\partial}{\partial y}\left(\varepsilon_s h \frac{\partial C_k}{\partial y}\right) + \alpha_s \omega_{sk}(C_{*k} - C_k) \tag{10}$$

Bed load:

$$\frac{\partial(\delta_b \bar{c}_{bk})}{\partial t} + \frac{\partial(\alpha_{bx} q_{bkx})}{\partial x} + \frac{\partial(\alpha_{by} q_{bky})}{\partial y} + \frac{1}{L_t}(q_{bk} - q_{b*k}) = 0 \tag{11}$$

Bed changes:

$$(1-p')\frac{\partial Z_{bk}}{\partial t} = \alpha\omega_{sk}(C_k - C_{*k}) + (q_{bk} - q_{b*k})/L_t \tag{12}$$

where ε_s is the eddy diffusivity of sediment in the vertical z direction; C_k is the concentration of the k-th size class, and C_{*k} is the corresponding transport capacity; α_s is the adaptation coefficient for suspended load; ω_{sk} is the sediment settling velocity; δ_b is the thickness of bed layer; c_{bk} is sediment load of the k-th size class; q_{b^*k}, q_{*k}, q_{bxk} and q_{byk} are the bed load transport capacity, the bed load transport rate, and transport rate components in x and

y directions, respectively; L_t is the adaptation length for bed load that characterizes the distance for a sediment process adjusting from a non-equilibrium state to an equilibrium state, and the model computes the non-equilibrium sediment transport process including the actual and equilibrium sediment load by considering the adaption length and factors; p' is the porosity of bed material, and Z_{bk} is the bed change.

In most of the early sediment transport models, it is usually assumed that the actual sediment transport rate or a saturated concentration is equal to the capacity of flow carrying sediment at equilibrium conditions. Based on this sediment transport state, a sediment transport model is called an equilibrium transport model. It could be a steady exchange of particles between bed material and the sediment load. However, this equilibrium assumption is not realistic in natural alluvial rivers because of variations in flow conditions and bed material properties. Non-equilibrium sediment transport models adapt sediment transport equations to determine the realistic bed load and suspended load transport rates.

In the processes of vertical integration, the water must be assumed to be shallow relative to channel width and the vertical variation in the flow to be negligible; otherwise, the dispersion terms must be maintained to preserve the effect of vertical sediment profiles on changes in the bed form. In the second case, the source term is nonzero but represents the dispersion terms. Computation of the dispersion term is complicated and requires knowledge of the vertical velocity and suspended sediment profiles in the x and y directions. The dispersion terms in suspended load transport equation are usually combined with the diffusion terms. More details can be found in the technical report of the sediment module [2].

3. Bed Geometry Adjustment Mechanism for a Local Scour Hole

Figure 3 displays the modules and flowchart of the calculation process in this study. First, the flow module calculates the flow conditions, e.g., flow field, velocity, and water depth. After calculating the flow conditions, the sediment transport module calculates the sediment transport process, e.g., sediment transport capacity, bed material sorting, and bed changes. Finally, local scour holes around hydraulic structures are adjusted using the bed geometry adjustment mechanism proposed in this study.

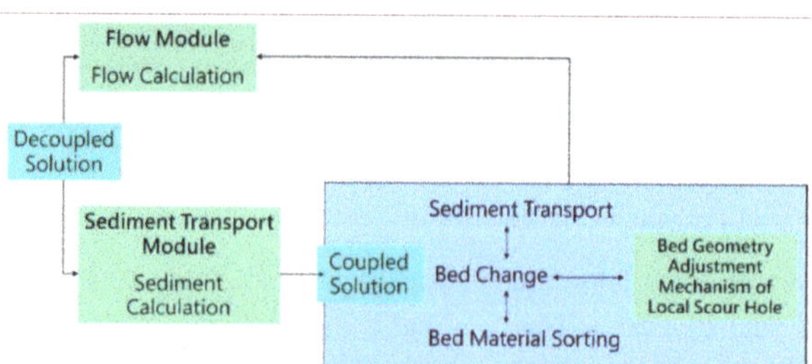

Figure 3. Flowchart of the computational modules used in this study.

The flow velocity profile in the vertical direction is generally non-uniform. The predicted local scour depth and volume are often underestimated due to the hypothesis of the depth-averaged velocity in the 2D model. In this study, an empirical formula for calculating the sediment repose angle [16] and an algorithm for modifying the bed geometry in the local scour hole are integrated into the 2D mobile-bed model to improve the model's ability to simulate local scours around the structures. Experiments were conducted by Meng et al. [16] by using cohesionless heavy sediment to obtain the angle of repose by using the natural accumulation method and angle of surface friction by using the tilting method. The experimental results revealed that the angle of repose over the water is bigger than that in the submarine condition. Moreover, the angle of repose does not increase monotonously with an increase in the sorting coefficient but increases with a decrease in the asymmetry coefficient S of non-uniform sediment for the same median diameter D_{50}.

The empirical formula (expressed in Equations (13)–(15)) proposed by Meng et al. [16] for uniform and non-uniform sediment is employed in this study. The procedure of the bed geometry adjustment mechanism for local scour holes can be explained as follows.

First, the angle of repose is calculated according to the sorting coefficient and bed material (e.g., uniform sand, non-uniform sand, gravel, or cobble). Then, the bed elevation around the hydraulic structures in the stream is calculated and modified using trigonometry with different bed slope trends (e.g., downhill or uphill slope). Finally, the adjusted bed geometry of the local scour hole around the hydraulic structures is obtained.

The angle of repose for uniform and non-uniform sediment proposed by Meng et al. [16] can be expressed as follows:

$$\phi_i = 6.08D_i + 30 \tag{13}$$

$$\phi_i = (3.05D_i + 35.7) + (3.05D_i + 35.7) \times \left[-0.41(S_{0i} - 1)^2 + 0.21(S_{0i} - 1) - 0.06\right] \tag{14}$$

$$S_0 = \sqrt{D_{75}/D_{25}} \tag{15}$$

where ϕ_i is the angle of repose at computational node i and D_i is the mean diameter of the bed material at computational node i. Moreover, S_0 is the sorting coefficient, in which D_{75} and D_{25} are the particle sizes at 75% and 25% weight of the bed materials. The average angle of repose between computational nodes i and $i + 1$ can be expressed as follows:

$$\phi = (\phi_i + \phi_{i+1})/2 \tag{16}$$

Depending on whether there is a downhill or uphill bed slope around the scour hole (Figure 4), the adjusted bed geometry can be expressed as follows:

$$Z_i' = \Delta x \cdot \tan\phi + Z_{i+1} \tag{17}$$

$$Z_{i+1}' = \Delta x \cdot \tan\phi + Z_i \tag{18}$$

where Z_i' is the bed elevation at computational node i that is modified by the angle of repose and Δx is the distance between computational nodes i and $i + 1$. In the computational nodes of bed elevation in the 2D model, where the set of i are the nodes along the longitudinal direction (x-direction) and the set of j are those along the lateral direction (y-direction). Figure 5 shows a flowchart of the bed geometry adjustment mechanism for a local scour hole. According to the characteristics of sediment classes, the bed geometry of local scour hole could be adjusted in the 2D model by considering the adjustment mechanism mentioned above.

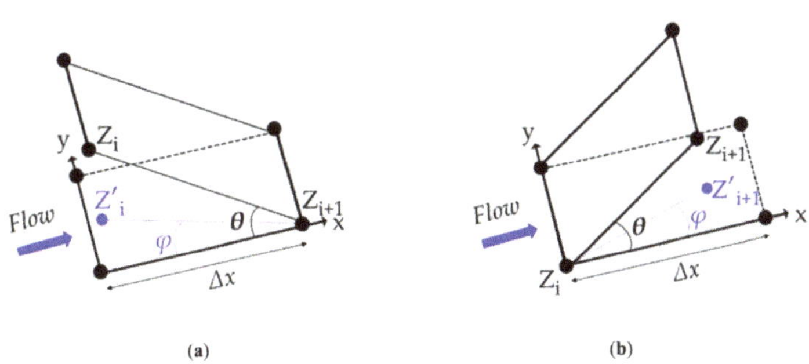

Figure 4. Different situations of bed elevation in the scour hole: (**a**) downhill bed slope ($S > 0$) and (**b**) uphill bed slope ($S < 0$).

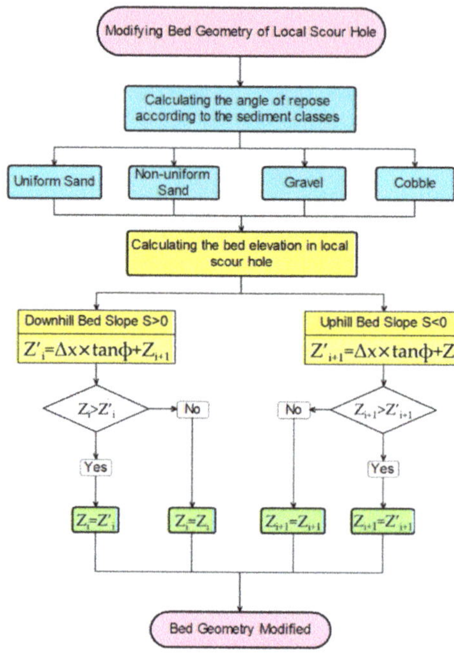

Figure 5. Flowchart of the bed geometry adjustment mechanism of a local scour hole.

4. Case Study

4.1. Initial and Boundary Conditions for the Experimental Case

An experimental case of multiple spur dikes is used for model calibration and validation. Figure 6 displays an overview of the experimental flume and spur dikes. The flume is 10 m long and 0.4 m wide with five spur dikes in the middle reach. The spur dikes are 0.1 m long and 0.02 m wide and are placed at a mutual distance of 0.2 m. The bed elevation and morphological change are obtained using the laser rangefinder system established on the flume. The terrain is measured at every 1-cm in the longitudinal and lateral directions of the experimental domain.

Table 1 lists the initial and boundary conditions of the experimental cases used for model calibration (Case No. G4a) and model validation (Case No. G5a). The inflow is steady with clear water. Figure 7 illustrates the computational domain of the experimental case. The grid numbers along the x-direction and y-direction are 337 and 63, respectively. The output profiles of the longitudinal bed elevation, e.g., A-line (Y = 0.344 m), B-line (Y = 0.252 m), C-line (Y = 0.196 m), D-line (Y = 0.146 m), and E-line (Y = 0.084 m) are presented. Three size classes ranging from 0.149 to 0.84 mm are selected as the bed material to compute the associated sediment transport.

Table 1. Initial and boundary conditions for the experimental cases used in model calibration and validation.

Case No.	Q (m^3/s)	V (m/s)	h (m)	F_r
G4a	0.0018	0.166	0.027	0.32
G5a	0.0023	0.17	0.033	0.30

Note: Q = discharge, V = averaged velocity, h = averaged water depth, F_r = Froude number.

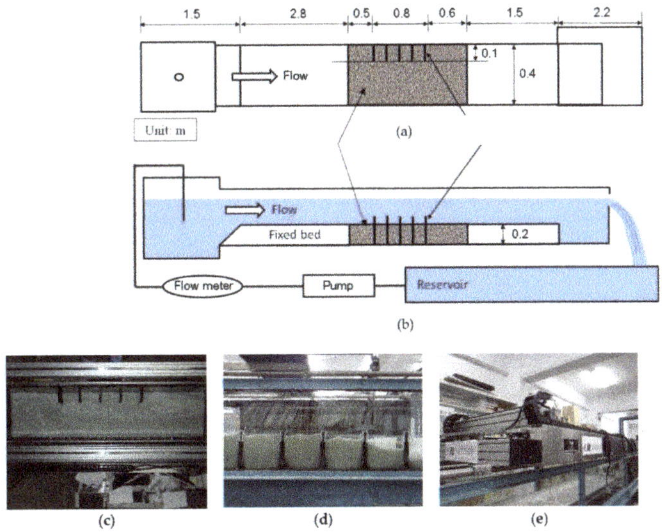

Figure 6. Overview of the multiple spur dike experimental channel: (**a**) top view, (**b**) side view, (**c**) top view of the morphological changes after conducting the experiment, (**d**) side view of the morphological changes after conducting the experiment, and (**e**) laser rangefinder system.

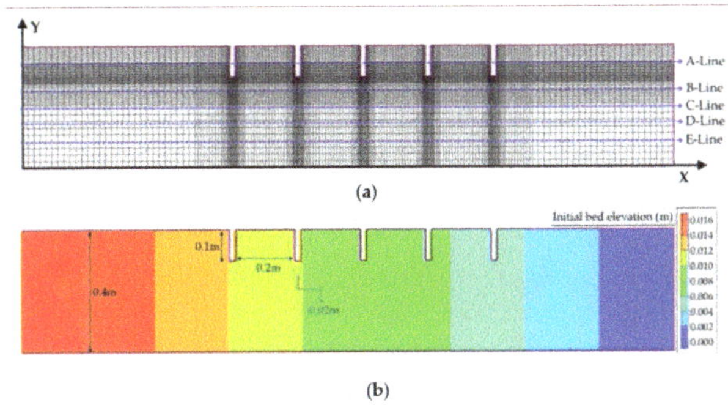

Figure 7. Computational domain of the experimental case: (**a**) computational mesh (63 × 337 grid) and (**b**) initial bed elevation and layout of multiple spur dikes.

4.2. Parameters for the Case Study

Table 2 shows the parameters selected in the model calibration. In the computational mesh, the total grids are 21,231 where the minimum lateral and longitudinal mesh sizes around the spur dikes are 0.5 cm. In the flow module, the computational time step is 0.1 s, and the equilibrium total simulation time is about 18,000 s. The mixing length eddy viscosity model is selected. In the case of a spur dike, the distance to the wall should be used as the length scale instead of that to the bed. Otherwise, the depth integrated coefficients for the eddy viscosity would be too large when the interior nodes are close to the wall. According to the CCHE2D technical report, the turbulence viscosity coefficient is a coefficient to adjust the value of the eddy viscosity. Its default value is set to 1 and it can be adjusted by users from 1 to 10. Another important issue regarding the mixing length model is the wall effect. Very close to the wall, the distance to the wall should be used as the length scale. Otherwise, the depth integrated coefficients for the eddy viscosity would be too large when the interior nodes are close to the wall. In CCHE2D model, the normal distance from a node to the wall is used to calculate the mixing length. It is also used to calculate the parabolic profile. This approach avoids the prediction of very large eddy viscosity near the wall. More details of the turbulence model and wall effect modification can be obtained in the CCHE2D technical report [2].

Table 2. Parameters used in model calibration.

	Parameters	Values
Computational mesh	Total grids (lateral × longitudinal)	63 × 337
	Minimum lateral mesh size (cm)	0.5
	Maximum lateral mesh size (cm)	2.0
	Minimum longitudinal mesh size (cm)	0.5
	Maximum longitudinal mesh size (cm)	2.0
Flow module	Time step (s)	0.1
	Total simulation time (s)	18,000
	Turbulence model	Mixing length model
	Turbulence viscosity coefficient	2
	Wall slipness coefficient	1
Sediment transport module	Sediment transport formula	Wu et al. [21]
	Minimum mixing layer thickness (m)	0.01
	Adaptation length for bed load transport (m), L_t	0.4
	Adaptation factor for suspended load, α_s	0.5
	Bed material size class 1 (mm)	0.149
	Bed material size class 2 (mm)	0.42
	Bed material size class 3 (mm)	0.84

In the sediment transport module, the sediment transport formula of Wu et al. [21] is selected for the model calibration. It related the bed-load transport rate to the non-dimensional excess grain shear stress with critical shear stress for incipient motion and grain shear stress. The variation of bed material gradation in the mixing layer can be described by the partial differential equation, e.g., Equations (10) and (11), while in other layers under the mixing layer, the bed material gradation can be determined by using the mass conservation law. The bed material is often divided into multiple layers, and the top layer is the mixing layer. The variation of bed material gradation is subject to exchange with those moving with flow. The mixing layer thickness is related to sand dune height or particle diameters of bed material. The minimum mixing layer thickness is set to 0.01 m in this study.

To calculate the non-equilibrium sediment transport, the adaptation length for bed load and adaptation factor for suspended load are considered, respectively, in the model. In general, the adaptation length characterizes the distance for sediment to adjust from a non-equilibrium state to an equilibrium state. It is a length scale for the river bed to respond to the disturbance of the environment, such as hydraulic structures, channel geometry changes and incoming sediment variation. The adaptation factor for suspended load is a use-specified parameter which is related to water depth, bed shear velocity, and settling velocity of sediment particles. In this study, the adaptation length for bed load transport is set to 0.4 m, and the adaptation factor for suspended load is set to 0.5 which have the best results. Figure 8 shows the bed material grain size distribution of the experimental case. The median diameter is 0.42 m, and the specific gravity is 2.5. Three bed material size classes are selected in the model based on the measured data of experimental case.

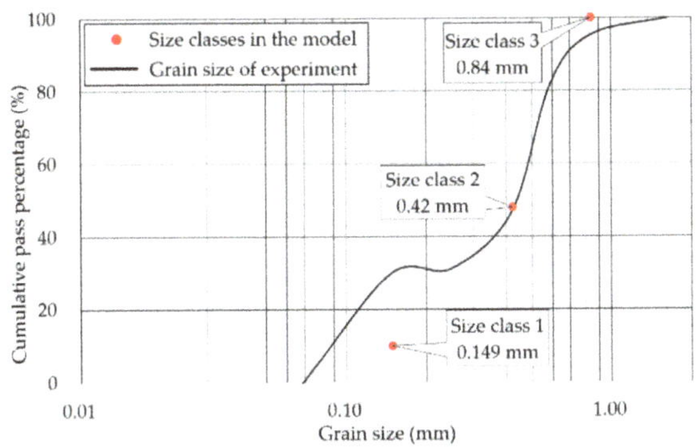

Figure 8. Bed material grain size distribution of the experimental case.

4.3. Flowchart of Model Calibration and Validation

The selection of flow and sediment parameters in the model calibration and validation are based on the sensitivity test. The sensitive parameters would be selected for the model calibration. Figure 9 shows the flowchart of model calibration and validation using the original model. First, it is needed to setup the model parameters, initial bed elevation, initial water depth, and boundary conditions. Then, the steady flow conditions on fixed bed are calculated and compared with the measured water stages. Under the standard of error analysis, the time step, turbulence model, turbulence viscosity coefficient, and wall slipness coefficient of flow module should be adjusted and calibrated by comparing the simulated results with the measured data. After the calibration of flow module, the sediment transport module on mobile-bed would be computed and compared with the measured bed changes. The sediment parameters (e.g., sediment transport formula, minimum mixing layer thickness, adaptation length for bed load transport, adaptation factor for suspended load) should be adjusted and calibrated. Finally, to complete the model validation, another set of the same experimental case is carried out based on the same model parameters.

5. Results

5.1. Simulation Results for the Experimental Case, Obtained Using the Original Model

Figures 10 and 11 show an overview of the simulated flow fields and velocity magnitude around the multiple spur dikes. The whole velocity magnitude of case G5a is higher than case G4a. Due to the setup of spur dikes in the experiment, there are contraction effects near the region, and these effects cause the flow velocity to increase. Moreover, eddy areas are present behind the dikes. The reattachment point shows that the divided streamline reattaches to the dike. When the flow passes through multiple spur dikes, a uniform flow is gradually obtained because of the increased width of the channel. Parts of the water flow into the spur dike field then make the circulation. The transported sediment causes deposition in this region around the dikes.

Figures 12 and 13 show the simulated and measured non-dimensional bed changes around multiple spur dikes in the original model when the equilibrium time is about 18,000 s. These changes are divided on the basis of water depth to compare the simulated and measured non-dimensional results. There are deposition bars occurring in the fields of multiple spur dikes, and the erosion occurs around the head of the first dike and the last dike. There is general erosion occurring along the longitudinal direction of the experimental flume. The location and magnitude of maximum erosion depth are predicted accurately in both experimental simulations. The calculated depths are 2.03 and 4.03 cm for the G4a and G5a cases. Moreover, the measured values are 2.00 and 4.19 cm. However, the angle of repose and modified bed geometry for the scour hole are not yet considered, and the predicted scour location is biased.

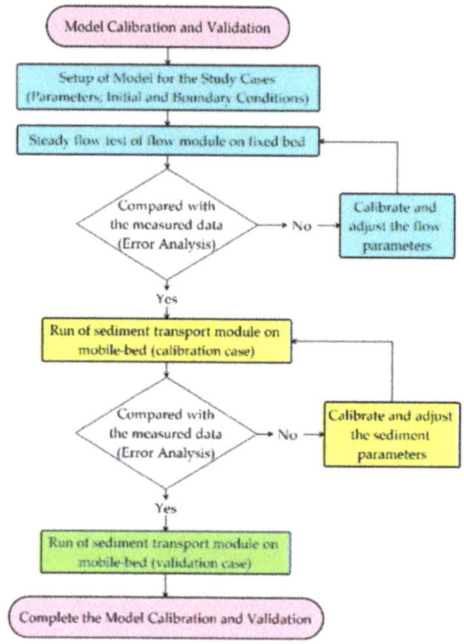

Figure 9. Flowchart of model calibration and validation using the original model.

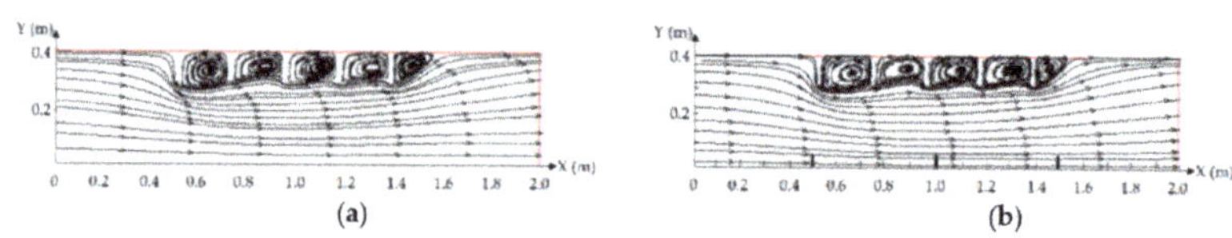

Figure 10. Overview of the simulated flow fields around multiple spur dikes in the experimental case: (**a**) case G4a and (**b**) case G5a.

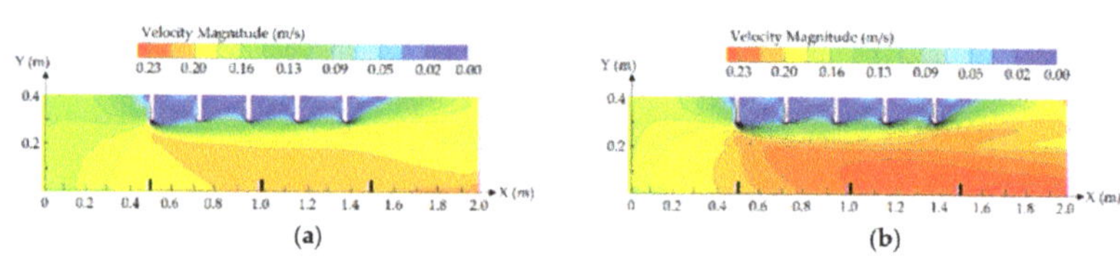

Figure 11. Overview of the simulated velocity magnitude around multiple spur dikes in the experimental case: (**a**) case G4a and (**b**) case G5a.

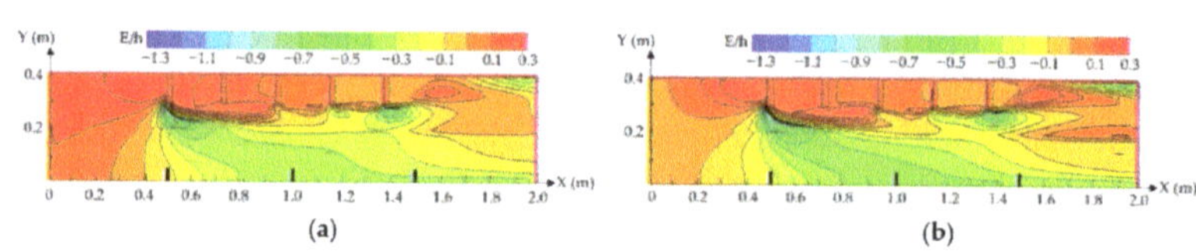

Figure 12. Overview of the simulated non-dimensional bed changes around multiple spur dikes: (**a**) case G4a and (**b**) case G5a. *E* is the scour depth, and *h* is the average water depth.

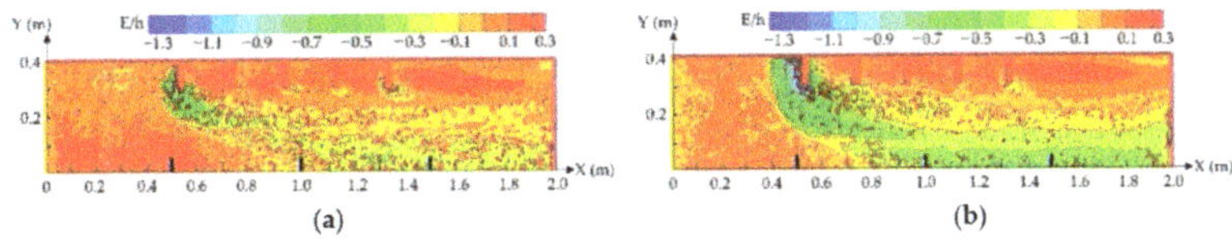

Figure 13. Overview of the measured non-dimensional bed changes around multiple spur dikes: (**a**) case G4a and (**b**) case G5a. E is the scour depth, and h is the average water depth.

5.2. Simulated Bed Changes in the Original and Adjusted Model

Based on the parameters of model calibration and validation, the improved bed geometry adjustment is considered and compared with the original model. Figure 14 displays an overview of the simulated non-dimensional bed changes around the multiple spur dikes when considering the angle of repose and bed geometry adjustment mechanism proposed in this study. According to the simulation results obtained using the adjusted model, significant local scouring is observed around the head of the first spur dike. The scour hole shape around the first spur dike could be simulated more obviously.

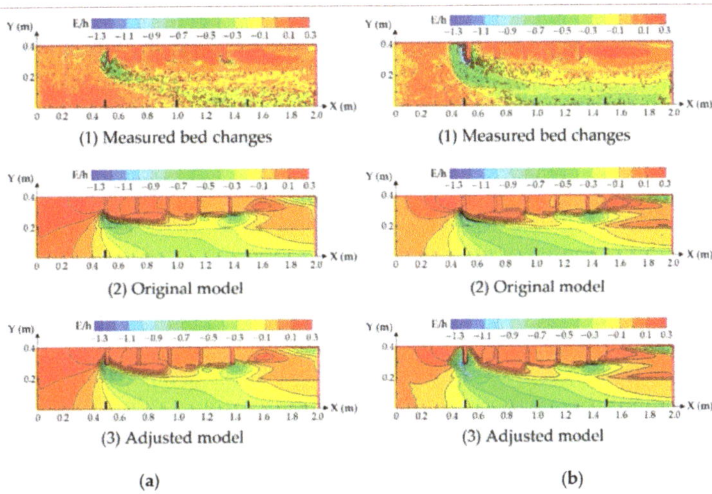

Figure 14. Overview of the simulated non-dimensional bed changes around multiple spur dikes when considering the angle of repose and adjusted bed geometry: (**a**) case G4a and (**b**) case G5a.

Figure 15 presents a local view and comparison of the simulated results obtained using the original and adjusted models. In the original model, head scouring is found to only occur near the first spur dike. The local scour holes around the spur dikes and the erosion depth are predicted more accurately when using the proposed mechanism.

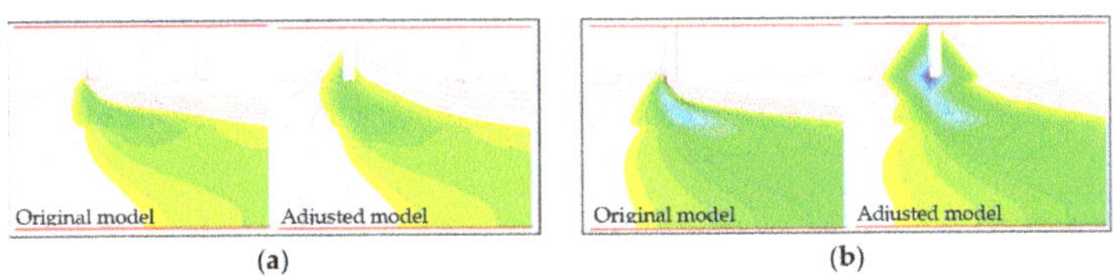

Figure 15. Local view of the simulated non-dimensional bed changes around the spur dike: original and adjusted model in (**a**) case G4a and (**b**) case G5a.

6. Discussion and Conclusions

6.1. Discussion

Figure 16 displays a sketch of a local scour hole around a spur dike; the sketch is used to compare the measured and simulated local scour hole shape [22]. Here, L is the distance of a vertical spur dike to a flume wall, and L_{sa}, L_{sb}, and L_{sc} are the distance from the head of the spur dike to the upstream, lateral, and downstream sides of the scour hole. The non-dimensional distance is calculated to compare the modeling and measured scour hole shape results. Table 3 presents a comparison of the non-dimensional distance of the scour hole around the spur dike. The simulated results reveal that the modified model makes 21.43% more accurate predictions than the original model, especially for the non-dimensional distance of L_{sa}/L in case G4a. This is discovered because the sediment repose angle and bed geometry adjustment mechanism are considered in the model. The simulated non-dimensional distances from the head of the spur dike to the lateral and downstream sides of the scour hole are in agreement with the measured data. The proposed mechanism provides an improved modification for 2D simulation of the area around hydraulic structures.

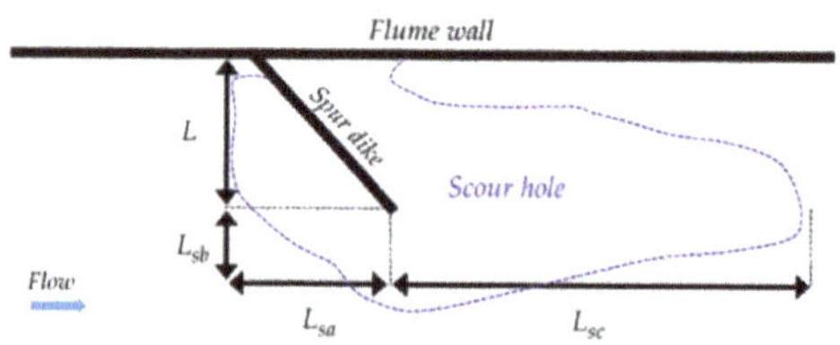

Figure 16. Sketch of a local scour hole around a spur dike. Modified from a study by Copeland [22].

Table 3. Comparison of the measured and simulated non-dimensional distance of the scour hole around the spur dike.

Case No.		Non-Dimensional Distance		
		L_{sa}/L	L_{sb}/L	L_{sc}/L
G4a	Measured	0.98	1.17	2.21
	Original Model	0.41	0.96	1.90
	Adjusted Model	0.62	1.15	1.92
	Improved (%)	21.43	16.24	0.90
G5a	Measured	1.34	1.54	3.12
	Original Model	0.78	1.02	2.16
	Adjusted Model	1.02	1.32	2.36
	Improved (%)	17.91	19.48	6.41

The longitudinal bed profiles in the experimental flume from A-Line to E-Line are compared with the original and adjusted modeling results to investigate the applicability and accuracy of the proposed bed geometry adjustment mechanism. It shows the relative scour trends around the spur dikes. The Mean Relative Error (MRE) and Root Mean Square Error (RMSE) are used for error analysis to evaluate the improvement obtained using the adjusted model. The MRE and RMSE can be expressed as follows:

$$MRE = \frac{1}{n}\sum_{i=1}^{n}\left|\frac{Z_i^{simu} - Z_i^{mea}}{Z_i^{mea}}\right| \tag{19}$$

$$RMSE = \sqrt{\frac{1}{n}\sum_{i=1}^{n}\left(Z_i^{simu} - Z_i^{mea}\right)^2} \tag{20}$$

where Z_i^{mea} and Z_i^{simu} are the measured and simulated bed changes at computational node i, respectively, and n is the total number of nodes measured along the longitudinal bed profile.

Figure 17 displays the measured and simulated bed elevation profiles along the longitudinal channel. The simulated results and measured data exhibit similar trends. There is a local scour near the head and body of the spur dikes. Due to the contraction effects on the hydraulic structures, general scour holes occur in the middle reach of the channel. The sediment particle size of the scour hole is gradually increased during the scouring process. Compared with the original model, the adjusted model determines a more accurate local scour hole. The improvement scope of adjusted model is decreasing from the head of spur dike to the other side of channel.

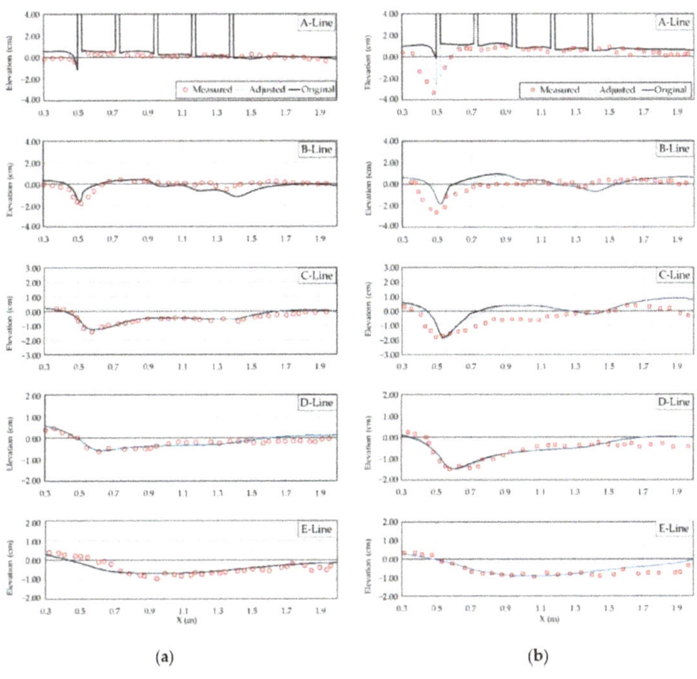

Figure 17. Comparison of the measured and simulated bed profiles from the A-line to the E-line: (**a**) case G4a and (**b**) case G5a.

Table 4 compares the MREs and RMSEs along the longitudinal bed profiles. In case G4a, the simulated bed elevations obtained using the adjusted model are in favorable agreement with the measured data along the A-line and B-line. Because of the increase in the inflow discharge of experiment, the scale of local scouring is more obvious in case G5a (Figure 18). The adjusted model provides a superior simulation of scour hole shape with more reasonable accuracy and reliability from the A-line to the D-line. The E-line has exceeded the scope of influence of local scour hole so that the original and adjusted model have the same trends.

Table 4. Comparison of the Mean Relative Errors (MREs) and Root Mean Square Errors (RMSEs) along the bed profiles from the A-line to the E-line.

Case No.		Bed Profiles				
		A-Line	B-Line	C-Line	D-Line	E-Line
G4a	MRE (cm)-Original	0.247	0.335	0.056	0.092	0.190
	MRE (cm)-Adjusted	0.226	0.319	0.056	0.092	0.190
G4a	RMSE (cm)-Original	0.255	0.422	0.123	0.125	0.240
	RMSE (cm)-Adjusted	0.244	0.403	0.123	0.125	0.240
G5a	MRE (cm)-Original	0.436	0.347	0.306	0.270	0.292
	MRE (cm)-Adjusted	0.195	0.296	0.284	0.268	0.292
G5a	RMSE (cm)-Original	0.776	0.370	0.368	0.379	0.322
	RMSE (cm)-Adjusted	0.347	0.351	0.342	0.313	0.322

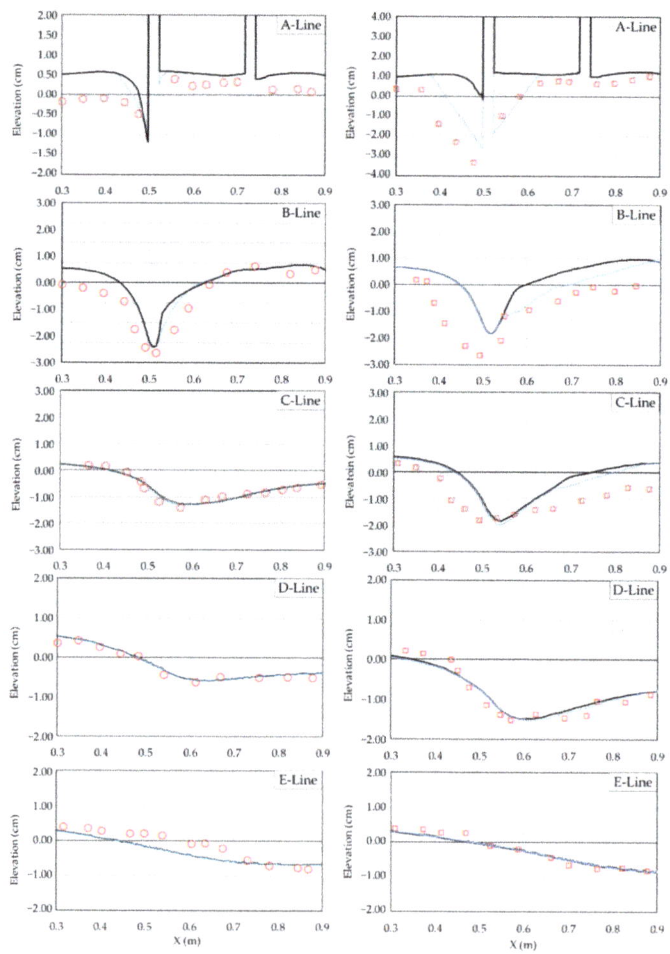

Figure 18. Local view of the comparison of the measured and simulated bed profiles at the A-line and E-line: (**a**) case G4a and (**b**) case G5a.

The empirical formulas proposed by Liu et al., Gill, and Neil [23–25], which are three common empirical formulas, are selected to calculate the head scouring depth of the first spur dike and compare the measured results with the modeling results. Table 5 presents a comparison of the empirical formula results and modeling results. The measured head scouring depth of the first spur dike is approximately 2 cm, and the depth in the modified model is approximately 2.25 cm. These values are in agreement with the measured data values. The original model calculates an underestimated head scouring depth of approximately 0.41 cm due to the simplification of 2D simulation around the structure. The calculation result obtained using each scour depth empirical formula has a large error compared with the measured scour depth. The proposed adjusted model can thus predict local scour around spur dikes with more reasonable accuracy.

Table 5. Comparison of the experiment, numerical model, and empirical formula results.

Measurement, Numerical Model, and Empirical Formula for Calculating the Head Scouring Depth		Scour Depth (cm)
Experimental Measurement	-	2.00
Original Model	-	0.41
Adjusted Model	-	2.25
Liu et al. [23] (1961)	$\frac{E}{h} = 2.15\left(\frac{L_p}{h}\right)^{0.4} F_r^{0.33}$	7.54
Gill [24] (1972)	$\frac{E+h}{h} = 8.375 \frac{(D_{50}/h)^{0.25}}{(1-L_p/B)^{0.857}}$	12.02
Neill [25] (1964)	$E = \alpha \left[\frac{2.5q^2}{D_{50}^{0.318}}\right]^{0.333}$	7.06–10.07

6.2. Conclusions

In the research of hydraulics and river dynamics, the understanding of the local scour phenomenon by the streams or the flood around hydraulic structures is important to reduce the risk of foundation structure failure. The empirical formulas are usually convenient to use for the calculation of maximum scour depth in the upstream or downstream of the structures. However, most of the empirical formulas have limitations to calculate the scouring depth or hole shape changing over time. This phenomenon and mechanism must be analyzed using the 2D or 3D numerical models.

In this study, a repose angle formula and bed geometry adjustment mechanism are integrated into a 2D mobile-bed model to improve the model's simulation accuracy for local scours around structures. The sediment classes of uniform sand, non-uniform sand, gravel, and cobble are considered in the proposed model. A multiple spur dike experimental case is simulated for the model calibration and validation. By adjusting the bed geometry of a scour hole depending on whether it is on a downhill or uphill slope, scour depth and geometry of local scour hole can be obtained that are in agreement with the measured situation. The Mean Relative Error (MRE) and Root Mean Square Error (RMSE) were calculated to compare the model results. In the validation case (G5a), the simulated results of the bed profile through the spur dike (A-Line) by adjusted model have a significant improvement effect. The improvement scope is decreasing from the head of spur dike to the side of channel.

According to the simulated results of the multiple spur dike experimental case, the sediment particle size of the scour hole is gradually increased during the scouring process. The scour hole side slope is steep at upstream of the spur dike, it is most affected by the sediment repose angle. In the comparison of the scour hole shape with the measured data, the modified model is 21.43% more accurate than the original model. A comparison of the calculated bed changes obtained using an empirical formula with measured bed change data reveals that the proposed numerical model, which includes the sediment repose angle and bed geometry adjustment mechanism, can predict local scours around spur dikes with more reasonable accuracy.

Author Contributions: Conceptualization, K.-C.Y. and C.-T.L.; methodology, Y.-C.L. and C.-T.L.; numerical simulation, Y.-C.L. and C.-T.L.; experiment, R.-K.J.; writing—original draft preparation, C.-T.L.; writing—review and editing, K.-C.Y., C.-T.L. and Y.J. All authors have read and agreed to the published version of the manuscript.

Funding: This research received no external funding.

Institutional Review Board Statement: Not applicable.

Informed Consent Statement: Not applicable.

Acknowledgments: The writers would like to acknowledge the Ministry of Science and Technology, Taiwan, that has supported this study. The Water Resources Agency, Ministry of Economic Affairs, Taiwan, has provided valuable field data and technical assistance.

Conflicts of Interest: The authors declare no conflict of interest.

Abbreviations

A_{xy}	turbulence viscosity coefficient
B	channel width
C	depth-integrated sediment concentration
C_k	concentration of the k-th size class
C_{*k}	concentration of the k-th size class corresponding transport capacity
C_s	integration constant
c_{bk}	sediment load of the k-th size class
D_{25}	particle size at which 25% by weight of the bed materials is finer
D_{50}	mean diameter
D_{75}	particle size at which 75% by weight of the bed materials is finer
D_i	mean diameter of bed material at computational node i
E	scour depth
f_{Cor}	Coriolis parameter
F_r	Froude number
g	gravitational acceleration
h	water depth
i	computational node
k	number of sediment size class
L	distance of spur dike vertical to flume wall
$\bar{l}$	depth integrated mixing length
L_p	length of dike
L_{sa}	distance from the head of spur dike to the upstream side of scour hole
L_{sb}	distance from the head of spur dike to the lateral side of scour hole
L_{sc}	distance from the head of spur dike to the downstream side of scour hole
L_t	adaptation length for bed load
n	total number of nodes along the longitudinal bed profile
p'	porosity of bed material
q	discharge per unit width
q_{b*k}	bed load transport capacity
q_{*k}	bed load transport rate
q_{bxk}	bed load transport rate components in x direction
q_{byk}	bed load transport rate components in y direction
Q	discharge
S	downhill or uphill bed slope between two nodes calculated by the bed geometry
t	time in governing equations
U	total velocity
u	depth-integrated velocity component in x direction
u_*	shear velocity
v	depth-integrated velocity component in y direction
V	averaged velocity
x	direction along the longitudinal bed profile
y	direction along the lateral bed profile
Z_{bk}	bed change

Z_i'	bed elevation at computational node i modified by the angle of repose
Z_i^{mea}	measured bed changes at computational node i
Z_i^{simu}	simulated bed changes at computational node i
α	coefficient of Neill (1980) [22]
α_s	adaptation coefficient for suspended load
β	coefficient to convert the turbulence eddy viscosity to eddy diffusivity for sediment
Δx	distance from computational node i to $i + 1$
δ_b	thickness of bed layer
ε_s	eddy diffusivity of sediment
κ	von Karman's constant
ρ	density of water
η	water surface elevation
θ	angle calculated based on the mesh nodes and the elevation of bed geometry
ς	relative depth of the flow
υ_t	eddy viscosity coefficient
τ_{xx}	depth integrated Reynolds stress in x direction at a plane perpendicular to x direction
τ_{xy}	depth integrated Reynolds stress in y direction at a plane perpendicular to x direction
τ_{yx}	depth integrated Reynolds stress in x direction at a plane perpendicular to y direction
τ_{yy}	depth integrated Reynolds stress in y direction at a plane perpendicular to y direction
τ_{bx}	shear stresses on the bed and flow interface in x direction
τ_{by}	shear stresses on the bed and flow interface in y direction
ϕ_i	angle of repose at computational node i
ω_{sk}	settling velocity of k-th size class sediment particle

References

1. Guney, M.S.; Aksoy, A.O.; Bombar, G. Experimental study of local scour versus time around circular bridge pier. In Proceedings of the 6th International Advanced Technologies Symposium (IATS'11), Elazığ, Turkey, 16–18 May 2011; Firat University: Elazver, Turkey, 2011.
2. Jia, Y.; Wang Sam, S.Y. *CCHE2D: Two-Dimensional Hydrodynamic and Sediment Transport Model for Unsteady Open Channel Flows Over Loose Bed*; Technical Report No. NCCHE-TR-2001-1; National Center for Computational Hydroscience and Engineering, The University of Mississippi: Oxford City, MS, USA, 2001.
3. Jia, Y.; Wang Sam, S.Y. *CCHE2D Verification and Validation Tests Documentation*; Technical Report No. NCCHE-TR-2001-2; National Center for Computational Hydroscience and Engineering, The University of Mississippi: Oxford City, MS, USA, 2001.
4. Lai, Y.G.; Greimann, B.P.; Wu, K. Soft bedrock erosion modeling with a two-dimensional depth-averaged model. *J. Hydraul. Eng.* **2011**, *137*, 804–814. [CrossRef]
5. Liao, C.T.; Yeh, K.C.; Huang, M.W. Development and application of 2-D mobile-bed model with bedrock river evolution mechanism. *J. Hydro-Environ. Res.* **2014**, *8*, 210–222. [CrossRef]
6. Juez, C.; Murillo, J.; García-Navarro, P. Numerical assessment of bed-load discharge formulations for transient flow in 1D and 2D situations. *J. Hydroinform.* **2013**, *15*, 1234–1257. [CrossRef]
7. Juez, C.; Murillo, J.; García-Navarro, P. A 2D weakly-coupled and efficient numerical model for transient shallow flow and movable bed. *Adv. Water Resour.* **2014**, *55*, 93–109. [CrossRef]
8. Juez, C.; Soares-Frazao, S.; Murillo, J.; García-Navarro, P. Experimental and numerical simulation of bed load transport over steep slopes. *J. Hydraul. Res.* **2017**, *55*, 455–469. [CrossRef]
9. Horvat, Z.; Isic, M.; Spasojevic, M. Two dimensional river flow and sediment transport model. *Environ. Fluid Mech.* **2015**, *15*, 595–625. [CrossRef]
10. Jia, Y.; Wang, S.S.Y. Simulation of horse-shoe vortex around a bridge pier. In Proceedings of the ASCE International Conference on Water Resource Engineering, Seattle, WA, USA, 8–12 August 1999.
11. Nagata, N.; Hosoda, T.; Nakato, T.; Muramoto, Y. Three-dimensional numerical model for flow and bed deformation around river hydraulic structures. *J. Hydraul. Eng.* **2005**, *131*, 1074–1087. [CrossRef]
12. Burkow, M.; Griebel, M. A full three dimensional numerical simulation of the sediment transport and the scouring at a rectangular obstacle. *Comput. Fluids* **2016**, *125*, 1–10. [CrossRef]
13. Castillo, L.G.; Carrillo, J.M. Scour, velocities and pressures evaluations produced by spillway and outlets of dam. *Water* **2016**, *8*, 68. [CrossRef]
14. Jia, Y.; Altinakar, M.; Guney, M.S. Three-dimensional numerical simulations of local scouring around bridge piers. *J. Hydraul. Res.* 2017. [CrossRef]
15. Yang, F.G.; Liu, X.N.; Yang, K.J.; Cao, S.Y. Study on the angle of repose of nonuniform sediment. *J. Hydrodyn.* **2009**, *21*, 685–691. [CrossRef]
16. Meng, Z.; Wang, H.; Yang, W. Experiment on angle of repose and angle of surface friction of cohesionless sediment. *J. Tianjin Univ. Sci. Technol.* **2015**, *48*, 11.

17. Al-Hashemi, H.M.B.; Al-Amoudi, O.S.B. A review on the angle of repose of granular materials. *Powder Technol.* **2018**, *330*, 297–417. [CrossRef]
18. Jhong, R.K. Study on Local Scour and Sediment Exchange around Spur Dikes. Ph.D. Thesis, Department of Civil Engineering, National Chiao Tung University, Hsinchu City, Taiwan, 2017.
19. Jia, Y.; Wang, S.S.Y. Numerical model for channel flow and morphological change studies. *J. Hydraul. Eng. ASCE* **1999**, *125*, 924–933. [CrossRef]
20. Jia, Y.; Wang, S.Y.Y.; Xu, Y. Validation and application of a 2D model to channels with complex geometry. *Int. J. Comput. Eng. Sci.* **2002**, *3*, 57–71. [CrossRef]
21. Wu, W.M. *CCHE2D Sediment Transport Model*; Technical Report No. NCCHE-TR-2001-3; National Center for Computational Hydroscience and Engineering, The University of Mississippi: Oxford City, MS, USA, 2001.
22. Copeland, R.R. *Bank Protection Techniques Using Spur Dikes*; Hydraulics Laboratory, U.S. Army Engineer Waterways Experiment Station: Vicksburg, MS, USA, 1983.
23. Liu, H.K.; Chang, F.M.; Skinner, M.M. Effect of bridge construction on scour and backwater. In *CER 60 HKL 22*; Colorado State University Civil and Environmental Engineering Section: Fort Collins, CO, USA, 1961.
24. Gill, M.A. Erosion of sand beds around spur dikes. *J. Hydraul. Div.* **1972**, *98*, 1587–1602. [CrossRef]
25. Neill, C.R. *River-Bed Scour—A Review for Engineers*; Technical Publication No. 23; Canadian Good Roads Association: Ottawa, ON, Canada, 1964.

water

MDPI

Article

Experimental Study on the Influence of the Transition Section in the Middle of a Continuous Bend on the Correlation of Flow Movement in the Front and Back Bends

Kanghe Zhang, Li Chen *, Yuchen Li, Bowen Yu and Yule Wang

State Key Laboratory of Water Resources and Hydropower Engineering Science, Wuhan University, Wuhan 430072, China; zkh@whu.edu.cn (K.Z.); lyc2016@whu.edu.cn (Y.L.); yubowen5@126.com (B.Y.); ylwang94@whu.edu.cn (Y.W.)

* Correspondence: chenliwuhee@whu.edu.cn

Received: 27 September 2020; Accepted: 14 November 2020; Published: 17 November 2020

Abstract: There exists a correlation in the flow movement between the front and back bends of a continuous bend, and the change in the transition section configuration influences this correlation. In this paper, laboratory experiments were conducted to systematically measure the three-dimensional velocity in a continuous bend with different width/depth ratios of the transition section. Based on this work, the flow movement characteristics of a continuous bend were analysed, including the circulation structure, circulation intensity and Turbulent Kinetic Energy (TKE). The flow movement correlation between the front and back bends of the continuous bend was also analysed. The influence of the width/depth ratio of the transition section on the correlation of the flow movement of the front and back bends and their relationship with discharge were explored. This research could help to elucidate the development and evolution laws of the continuous bend and provide theoretical support for flow movement, flood routing, sediment transport and riverbed evolution. It is found that in addition to the circulation structure and intensity, the TKE of the front and back bends of the continuous bend also shows a strong correlation. With increasing discharge, the correlation between the front and back bends increases, and the larger the discharge is, the greater the influence of the same amplitude of variation in the discharge on the correlation. At the same time, the larger the discharge is, the greater the influence of the same amplitude of variation in the width/depth ratio of the transition section on the correlation of the front and back bends. When the discharge is constant, the correlation between the front and back bends decreases with the decrease in the width/depth ratio of the transition section, and the smaller the width/depth ratio, the greater the influence of the same amplitude of variation in the width/depth ratio on the correlation. There is no linear relationship between the amplitudes of variation in both the discharge and width/depth ratio and the corresponding amplitude of variation in the correlation. With increasing discharge, the amplitude of variation in the correlation caused by the same amplitude of the variation in discharge slightly increases. However, there is an exponential relationship between the amplitude of variation in the width/depth ratio and the corresponding correlation. The influence of the width/depth ratio on the correlation is clearly greater than that of the discharge.

Keywords: continuous bend; transition section; correlation of flow movement; width/depth ratio; circulation structure; circulation intensity; turbulent kinetic energy

---

1. Introduction

There are many continuous bends in nature; see Figure 1. Unlike single bend, the flow characteristics of any bend along a continuous bend are not only related to its own configuration and the composition of the riverbed boundary but also influenced by the upstream and downstream reaches [1,2]. In a continuous bend, a front bend, back bend and transition section function as a whole, where the front and back bends are correlated with each other through the transition section, making the flow movement characteristics in the continuous bend more complex, especially in the back bend. As early as the 1930s, Mockmore [3] noticed that the flow characteristics of the back bend in a continuous bend are influenced by the Residual Circulation of the Front Bend (RCFB). Later, both numerical simulations [4,5] and laboratory experiments [6,7] pointed out that circulation produced in the front bend could exist in the back bend, influencing flow structure such as velocity distribution [8], circulation structure [5,8–10] and turbulence parameters [9] in the back bend. Because of the above correlation between the "single bends" of the continuous bend, research on continuous bends is relatively difficult and has been one of the hotspots worldwide.

(a) (b)

Figure 1. Continuous bends in nature. (**a**) The Jingjiang Reach, Yangtze River, China. (**b**) The Mississippi River.

As the link between the front and back bends, the configuration of the transition section has an important influence on the correlation between the front and back bends of a continuous bend. At present, many scholars have carried out relevant researches. For example, from the perspective of flow characteristics, Ghanmi [11] investigated the variation in transverse velocity exiting the front bend in transition sections with different lengths by combining physical model and mathematical model experiments and pointed out that when the distance between the front and back bends is relatively small, the RCFB will extend to a certain distance in the back bend; when the length of the transition section is approximately 4R (R is the radius of the bend), the RCFB will disappear when the inlet section of the back bend is reached. If the length of the transition section continues to increase, the RCFB will completely disappear before entering the back bend. Liu [12] also pointed out that when the length of the transition section reaches 3~5 times the length of the bend reach, the influence of the front bend on the back bend is so small that it can be ignored. Hu et al. [13] analysed the position corresponding to complete attenuation and generation of the circulation structure, as well as the surface reflux in the back bend with different transition section lengths by means of mathematical model experiments and concluded that the longer the relative transition section length (L/R) is, the smaller the influence of the front bend on the back bend, but when the relative transition section length equals 0.375, the influence remains constant. Zhou et al. [14] and Zhou et al. [15] pointed out that if the length of the transition section is significantly larger than (less than) the complete attenuation length, the influence of the front bend circulation can be (cannot be) ignored. Jing and Zhao [16] studied the influence of the boundary configuration factor of the transition section of a continuous bend on the transverse gradient of the bend and found that the transverse gradient of the front bend and the transition section increased with the increase in the boundary configuration factor, while the back bend decreased in turn. From the perspective of sediment transport, Cai et al. [17] pointed out that to ensure good river channel conditions between the front and back bends and to avoid the appearance of sand ridges in the transition section, the length/width ratio of the transition section should be 1.5~4, with the most effective ratios being closer to 3. Wang et al. [18] pointed out that when the transition section between the two bends is too long or too short, a shoal easily forms in the transition section, and the most suitable length of the transition section corresponds to when the end of the circulation of the front bend can just reach the formation stage of the circulation in the back bend.

Previous studies have only investigated the influence of the length and boundary configuration of the transition section on the correlation between the front and back bends in a continuous bend [11,13,16,19], while there is a lack of studies on the influence of the width/depth ratio. The above studies focused on factors such as transverse velocity, circulation structure, circulation intensity, surface reflux and transverse slope and did not reveal the correlation of TKE and the relationship between the configuration of the transition section and the correlation. Moreover, the above studies were mostly mathematical model simulations and theoretical analyses, with regular cross-section configurations, which do not accurately represent natural riverbeds, and the reliability of the experimental results needs to be confirmed [20,21]. In summary, this paper adopts the compound cross-section flume experiment, uses an Acoustic Doppler Velocimeter (ADV) to measure the three-dimensional velocity of continuous bend at different discharges and different transition section width/depth ratios and obtains hydraulic elements that closely reflect an actual situation [22]. Based on the measured results, the correlations among the circulation structure, circulation intensity and TKE between the front and back bends of a continuous bend are analysed; moreover, the influence of the width/depth ratio of the transition section on the correlation between the front and back bends and its relationship with discharge is explored. According to the above studies, the importance of the transition section is revealed, and the flow characteristics in continuous bends are better understood.

The aims of this study were to explore the correlation between front and back bends of the continuous bend. Specifically, we will investigate the influence of the width/depth ratio of the transition section on the correlation of the flow movement of the front and back bends and their relationship with discharge. We will also compare the influence of discharge and width/depth ratio of transition section on the correlation between the front and back bends of the continuous bends, to help comprehensively understand the mechanics of flow in continuous river bends.

The above aims may help to elucidate the development and evolution laws of the continuous bend and provide theoretical support for a more comprehensive understanding of flow movement, flood routing, sediment transport and riverbed evolution in continuous bends and provide help for better numerical simulation of the continuous bend. Meanwhile, in engineering practice, a full understanding of the correlation of the front and back bends of the continuous bend and its relationship with discharge and transition section configuration can be used for more effective channel regulation and river management.

2. Experimental Setup and Methods

2.1. Design of the Experiment Flume

In this paper, 90° and 135° are selected as the central angles of the front and back bends of the continuous bend in the experimental flume. A 2 m long transition section is set between two bends, where the straight reach is tangent to the curved reach. The planform of the laboratory flume after generalization is shown in Figure 2a; a 7 m long straight reach is set up upstream and downstream of the continuous bend. The upstream straight reach ensures the stability of the flow entering the experiment reach, and the downstream straight reach reduces the influence of the tail gate on the flow structure. The radii of the central axes of the front and back bends are both 2.4 m, the width of the flume is 1.2 m, and the curvature ratio (r/B) is 2 [23]. If the curvature ratio is less than 2, it represents a sharp bend channel [24], so this flume can be regarded as a continuous sharp bend flume.

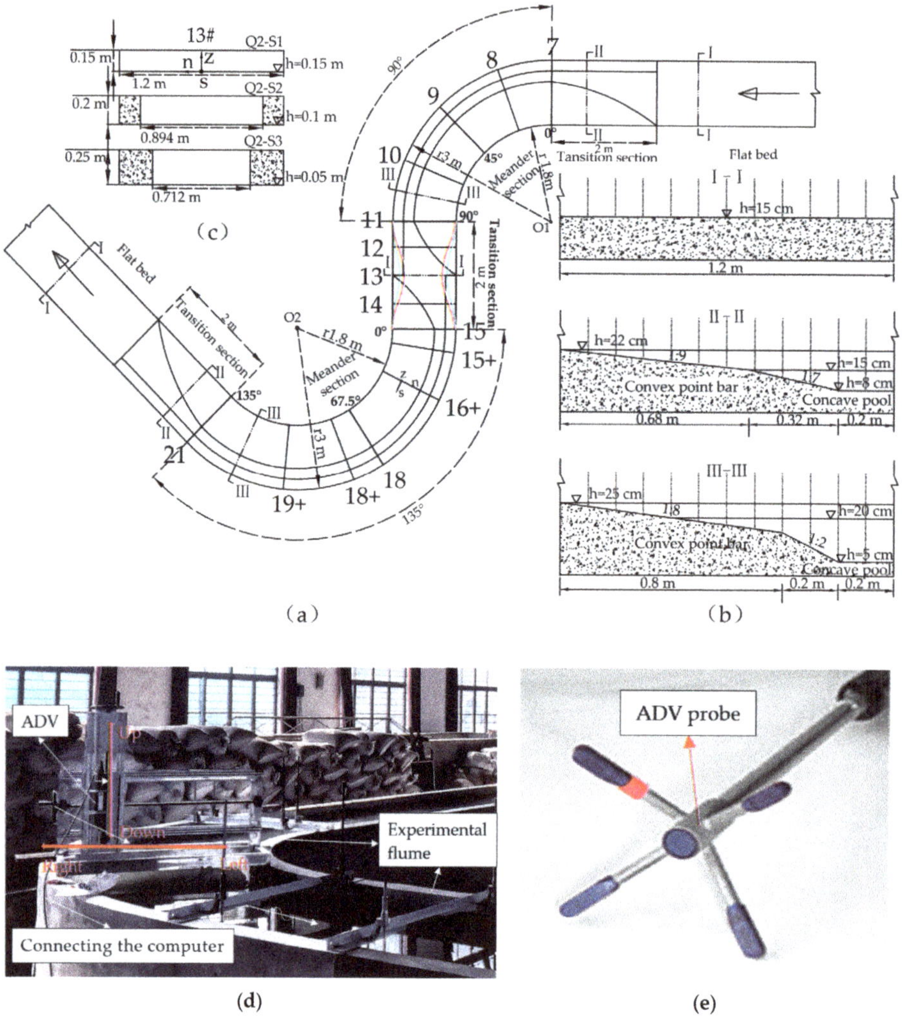

Figure 2. Design of the experimental flume. (**a**) Planform of the continuous bend. (**b**) Typical cross-section geometry. (**c**) Configuration of the section 13# at different scenarios. (**d**) Experimental setup on the model. (**e**) Acoustic Doppler Velocimeter (ADV) probe on the experimental setup.

In Figure 2b, III-III shows the cross-section geometry of the bend flume, in which the width of the point bar is 0.8 m, with a transverse slope of 1:8. The elevation of the bar top is 0.25 m, and that of the bar bottom is 0.15 m. The concave pool is composed of a transitional slope with a width of 0.2 m linking to the point bar, with a transverse slope of 1:2, and a 0.2 m wide flat pool, with a constant elevation of 0.05 m. The straight inflow and outflow reaches and section 13# in the transition section are designed to have a horizontal bed with an elevation of 0.15 m, and their cross-section geometry is shown in I-I in Figure 2b. At the downstream end of the inflow reach and the upstream end of the outflow reach, the transition section with a length of 2 m transforms smoothly from the horizontal bed to the bar-pool geometry. The geometry of a randomly selected cross-section in this transition section is shown in II-II in Figure 2b.

2.2. Experiment and Scenario Design

The selected discharges, corresponding water levels and various hydraulic parameters are shown in Table 1. The inflow discharge is controlled by an electromagnetic flowmeter with an accuracy of 0.5%, and the discharge is controlled by adjusting the valve; the tail door is a plate-turning type, and the water level of the flume is controlled by adjusting the opening degree of the tail door. The experimental flume is made of cement plastering, which is as smooth and uniform as possible to reduce the influence of the variability in roughness on flow.

Table 1. Experimental conditions and measured variables.

Condition	Discharge (l/s)	Water Stage (m)	Fr [3]	$\overline{R}/B$ [4]	B^*/H^* [5]	$\overline{B}/\overline{H}$ [6]	Re [7]
S1 [1]			0.096		9.6	9.84	8423
S2	Q1-16 [2]	0.275	0.078		5.11	6.6	9818
S3			0.067		3.16	5.67	10,510
S1			0.101		8.00	8.22	11,195
S2	Q2-22	0.3	0.088	2.0	4.47	5.5	12,978
S3			0.079		2.85	4.72	13,856
S1			0.127		6.86	7.05	17,237
S2	Q3-35	0.325	0.117		3.97	4.71	19,879
S3			0.109		2.59	4.04	21,170

[1] S denotes the scenario. [2] Q is discharge, 16, 22 and 35 are magnitude of discharges and denote 16 l/s, 22 l/s and 35 l/s, respectively. [3] Fr is the Froude number. [4] $\overline{R}$ is radius of the central line, B is the channel width of the continuous bend. [5] B^* is the width of section 13#, H^* is the flow depth of section 13#, B^*/H^* is the width/depth ratio of section 13#. [6] $\overline{B}$ is the mean channel width of the transition section, calculated by dividing plane area by total length of the transition section, where the plane area is obtained by the surface integral of the transition section boundary curve. $\overline{H}$ is the mean channel depth of the transition section, $\overline{B}/\overline{H}$ is the mean width/depth ratio of the transition section. [7] Re is the Reynolds number.

As shown in Figure 2a,c, section 13# has B1 = 1.20 m in the S1 state; in the S2 and S3 states, section 13# narrows to different degrees. After narrowing, the planform shapes of the transition sections are shown as cyan dotted lines and red dashed lines, and the corresponding widths are B2 = 0.89 m and B3 = 0.71 m, respectively. To ensure the consistency of the cross-sectional area, the transition sections of S2 and S3 are deepened accordingly.

The parameters of the three scenarios are shown in Table 1. After narrowing twice, the width/depth ratios of section 13# are significantly reduced, the variation amplitudes of Q1 are 47% and 67%, those of Q2 are 44% and 64%, and those of Q3 are 42% and 62%. The variation amplitudes of the width/depth ratio for the two scenarios at the three discharges gradually decrease with increasing discharge, but the differences are small. The variation amplitudes of the average width/depth ratio of the whole transition section for the two scenarios at the three discharges are consistent, 33% and 42.5%, respectively.

The calculation formulas of Fr and Re in the experiment are as follows:

$$Fr = \frac{U}{\sqrt{gH}} \tag{1}$$

$$\mathrm{Re} = \frac{UR}{\nu} \tag{2}$$

where U denotes average velocity of the cross-section, H represents average depth of the cross-section, g represents gravitational acceleration, R represents hydraulic radius, ν represents kinematic viscosity coefficient.

The Froude and Reynolds numbers of section 13# are chosen to represent the whole study area. Fr is calculated directly by Formula (1). By using the average width B^* and the average water depth H^* of the 13# section, the hydraulic radius R^* is obtained by the formula $R = BH/(B+2H)$, and Re is calculated by combining the average velocity U^* of the cross-section 13# and kinematic viscosity coefficient ν. The calculation results are shown in Table 1.

2.3. Data Collection and Processing

An Acoustic Doppler Velocimeter (ADV) was used to measure three-dimensional instantaneous velocity in the orthogonal curvilinear coordinate system. The position relationship between the measurement section and the orthogonal curvilinear coordinate system is shown in Figure 2a,c, where the downstream s axis coincides with the flume's centreline, the transverse n axis points in the left direction, and the vertical z axis is upwards. The accuracy of the ADV is 0.5%. The sampling frequency was 200 Hz, and the sampling time was more than 30 s, which ensured that there were at least 6000 original velocity data points at each measuring point. The experimental setup with ADV system is installed on the model (Figure 2d), keeping the probe downward and collecting the water velocity 5 cm below the probe. ADV probe is shown in Figure 2e. When measuring, keeping the four corners of the probe on the same horizontal plane, it is ensured that the red mark of the probe is in the same direction with the downstream flow. Because the sampling volume of the probe was located 5 cm from the sensors and the probe must be completely submerged in the water, the velocity in the range of 5 cm below the surface could not be measured by the probe; considering the accuracy of the data, extrapolation processing was not carried out in this paper.

Cross-section 13# was selected as the typical section, and the accuracy assessment tests had been conducted by validating integral quantity of the normal velocity measured with ADV on cross-section 13# against the discharge measured with electromagnetic flowmeter at different discharges and conditions. The results in Table 2 showed that the error rates were all within 5%, we could infer that velocity measured with ADV was accurate enough in this experiment.

Table 2. Instrument measurement error.

Discharge	Q1-S1	Q1-S2	Q1-S3	Q2-S1	Q2-S2	Q2-S3	Q3-S1	Q3-S2	Q3-S3
Electromagnetic flowmeter (Ls^{-1})	16	16	16	22	22	22	35	35	35
ADV (Ls^{-1})	15.6	16.4	16.6	22.17	22.1	22.7	34.3	34.94	35.67
Error (%)	2.50%	2.50%	3.75%	0.77%	0.45%	3.18%	2.00%	0.17%	1.91%

The concentration of particles in the water would influence the stability of the measurement signal. Generally, the Signal-to-Noise Ratio (SNR) must be larger than 15 to ensure data reliability [25]. When the SNR was less than 15, a small amount of plastic sand with a particle size smaller than 0.1 mm was added to the backwater system to increase the particle concentration. Both the particle size and density of the plastic sand were small, so although there was some settlement, the flow structure would not be affected.

The measurement interval of the front bend was 22.5° from section 0° to 90°, the measurement interval of the transition section was 0.5 m, and the measurement interval of the back bend was 33.75° from section 0° to 135°. The corresponding section number was shown in Figure 2a, and the additional sections 15+# and 18+# of the back bend, which were the generation and attenuation position of the circulation at some scenarios and discharges and located 11.25° downstream of sections 15# and 18#, were measured. Eleven vertical measuring lines were distributed at a horizontal interval of 10 cm across each measuring section, and 15 measuring points were arranged from bottom to top according to the intervals of $0.2h'$, $0.3h'$, $0.4h'$, $0.5h'$, $0.6h'$, $0.7h'$, $0.75h'$, $0.84h'$, $0.87h'$, $0.9h'$, $0.93h'$, $0.96h'$, $0.98h'$ and $1h'$, respectively (h' was water depth; increasing distribution density with increasing depth). Because the measurement points across the whole cross-section were relatively sparse, interpolation processing was carried out before the circulation structure, intensity and turbulence kinetic energy experimental results were presented. The inverse distance weight interpolation method was used for interpolation, and the cross-sectional distribution map was drawn using the interpolation results.

The ADV may have some peaks caused by Doppler noise and Doppler signal aliasing in the measured velocity series. These peaks may be due to the measured velocity exceeding the set range of measured velocity or affected by the reflection of the previously transmitted signal. All of these phenomena will lead to the distortion of velocity measurement [26,27]. Therefore, it is necessary to denoise the velocity time series, and the phase space threshold processing method with threshold change is used to identify peak values in the velocity series. Then, the points before and after the peak values are replaced by polynomials.

3. Results and Discussion

3.1. *Analysis of the Circulation Structure*

In bends, due to the combined action of centrifugal force and hydrostatic pressure, surface water flows to the concave bank, bottom water flows to the convex bank and projection on cross-section forms circulation. In the size of the circulation, the transverse or vertical velocity is always larger than zero.

In typical cross-sections, the time-averaged velocity vector field was synthesized by using the transverse and vertical time-averaged velocities, and the transverse circulation structure was obtained by connecting the vectors with the streamline. The circulation structure here represents the distribution shape and size of the circulation on the n, z coordinate plane. Because the flume adopts a complex cross-section shape that is closer to the actual river channel, the main circulation structure in the bend is mostly narrow, exhibiting an irregular ellipse as the cross-section configuration, which is different from regular rectangular or trapezoidal channels.

Figure 3 is a three-dimensional view of the variation in the circulation structure along the continuous bend at Q3. As shown in Figure 3, the streamline at the inlet section of the front bend begins to slightly bend, and there are signs of circulation generation. Then, a single clockwise transverse circulation is generated and gradually developed along the channel. At the outlet section, the circulation structure occupies the whole section. It can be seen from the red dotted line of the circulation centre position, the circulation centre gradually shifts from the bottom to the middle of the water body. In the transition section, we also observe a single clockwise circulation structure; however, the size is reduced, and the circulation centre is located closer to the left bank.

In the back bend channel, a single clockwise circulation can be seen at the inlet section. Since transverse circulation cannot be produced without centrifugal force, the clockwise circulation structure at the transition section and the inlet section are all the RCFB. The size of the RCFB decreases gradually along the channel, and the circulation structure is divided into two parts at the apex of the bend: the part moving to the surface of the concave bank and the part moving to the centre of the river channel. The RCFB near the concave bank is consumed by the Circulation of the Back Bend (CBB), approaching complete attenuation near 78.75° (purple dotted line), while the RCFB in the central area exists until close to the outlet of the back bend (red and orange dotted line). It should be noted that the very weak clockwise circulation structure near the concave bank gradually increases. Fariba and Alireza [28] found that the centrifugal force and cross-stream turbulent shear stress are main driving forces of near bank circulations and such conditions do exist near the concave bank. Combining the observed changes in the circulation structure along the channel, it can be inferred that this circulation is an "out-bank cell" (orange dotted line). After entering the back bend, an anticlockwise circulation structure begins to form near the bed of the concave bank (dark blue dotted line), and then an anticlockwise circulation structure forms above the convex point bar (blue dotted line). Along the channel, the two CBBs develop gradually, with the circulation in the concave pool being dominant. Afterwards, the two CBBs gradually merge together. In the back bend, a single circulation model is no longer applicable; two or more circulation structures with different vortex directions coexist, and the circulations in different directions squeeze each other, resulting in an irregular circulation structure.

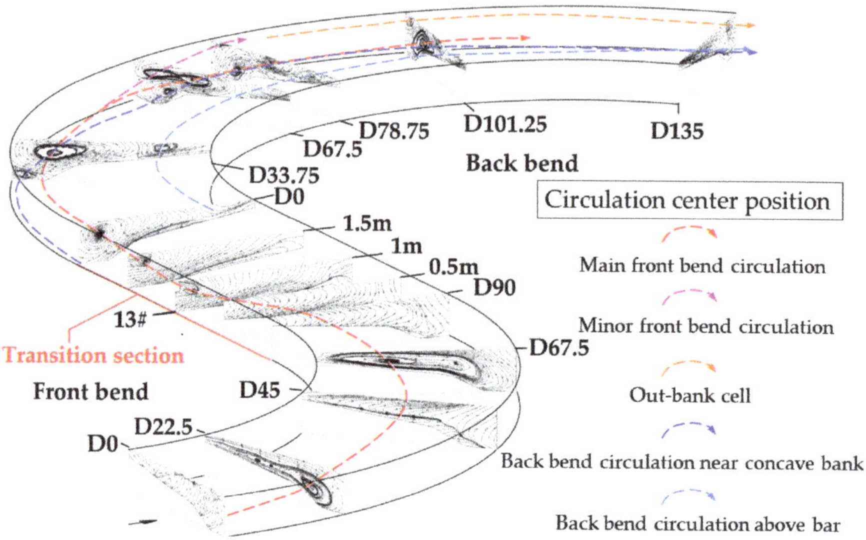

Figure 3. Distribution of circulation structures along the continuous bend at Q3. The direction of the arrow in the typical section is the flow direction. The red, purple, orange, dark blue and blue dotted lines denote different circulation centre positions along the continuous bend, respectively.

Typical sections are selected to analyse the influence of changes in the width/depth ratio on the circulation structure under different scenarios and discharges. As shown in Figure 4, with the decrease in the width/depth ratio of the transition section, at Q3, the circulation structure in the outlet section of the front bend changes little and is only slightly different in the upper part of the water body near the concave bank. However, at the three discharges, the circulation structure distribution at the apex section of the back bend all changes greatly. Especially at Q3, the size of the clockwise circulation structure decreases gradually and is divided into two parts, the component on the concave bank side gradually disappears; the size of the counterclockwise circulation structure gradually increases, and the two circulations gradually approach each other, nearly merging at S3. At other discharges, the RCFB at the apex section of the back bend decreases gradually, the CBB increases gradually.

To understand the change in the structure of the back bend more clearly, circulation structure distribution at typical cross-sections along the back bend in different scenarios is shown in Figure 5. Two or more circulation structures with different vortex directions coexist along the back bend in different scenarios. With the decrease in the width/depth ratio of the transition section, the size of the RCFB decreases gradually at the same cross-section, and minor RCFB rarely exists at cross-section 67.5° and either as main RCFB at cross-section 78.75° in S3. Meanwhile, the size of the CBB increased gradually at the same cross-section, and an anticlockwise circulation appears at the inlet section of the back bend at S3. There are no clockwise circulations near concave bank water surface at S2 and S3.

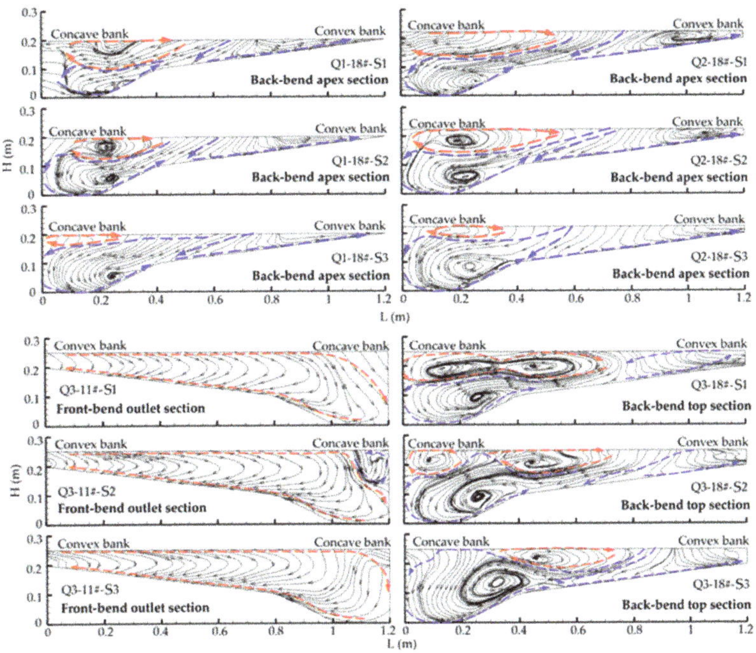

Figure 4. Circulation structure distribution of the typical cross-sections under different scenarios and discharges. The direction of the arrow in the typical section is the flow direction. The red and blue dotted lines denote the clockwise and anticlockwise circulation structure, respectively.

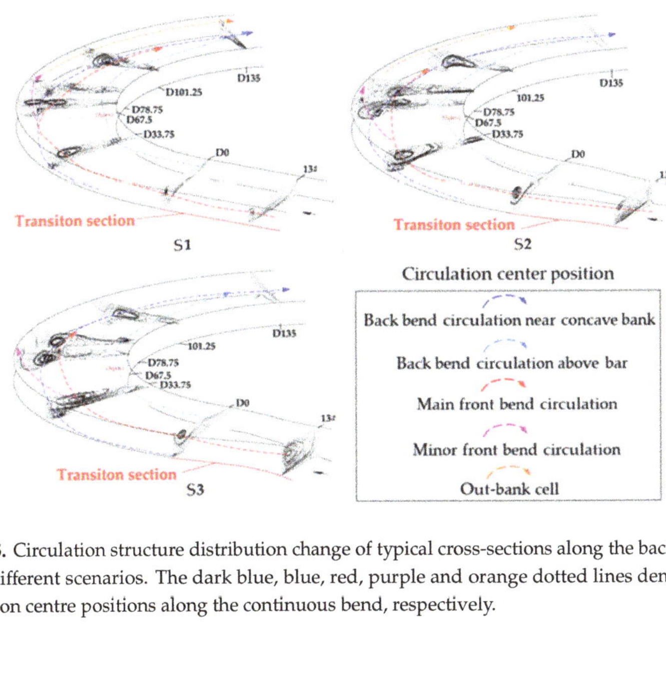

Figure 5. Circulation structure distribution change of typical cross-sections along the back bend at Q3 under different scenarios. The dark blue, blue, red, purple and orange dotted lines denote different circulation centre positions along the continuous bend, respectively.

The changes in the size of the two kinds of circulation structures along the channel are analysed for different discharges and different scenarios (the CBB includes two parts, one in the concave pool and another above the convex point bar). To perform a more accurate statistical analysis of the circulation structure size, a cross-section is divided into grids with a maximum spacing of 1 cm. After interpolation, velocity vectors are plotted in each grid to identify the range of circulation. The statistical size is the size within the outer contour of the cross-section circulation; if the grid is within the range of circulation, its size is added to the circulation structure size. The changes in the size of the two kinds of circulation structures along the channel are plotted, and then, the trend lines are fitted, as shown in Figure 6. The location information of the key changes in the two circulation structures are given in Table 3.

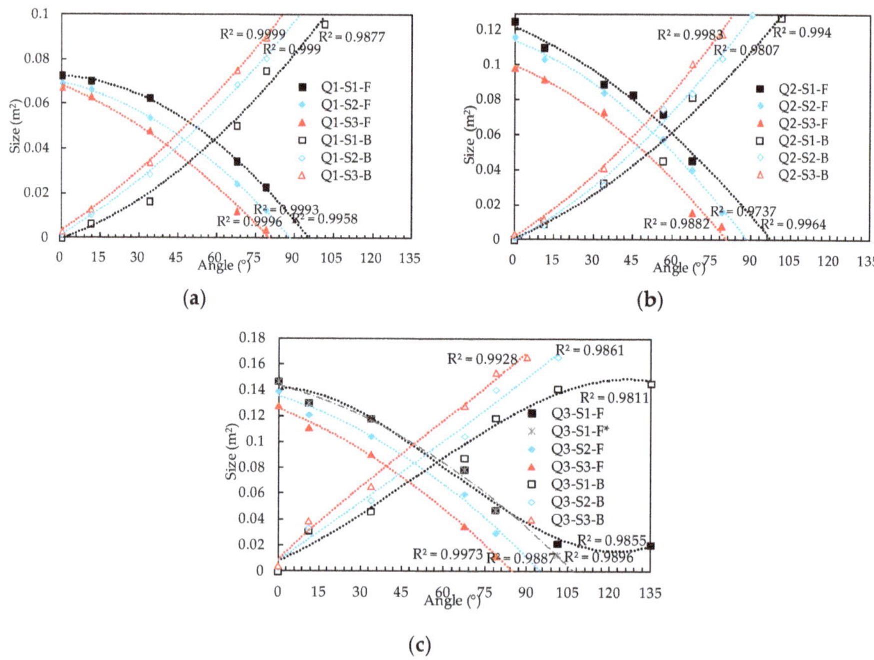

Figure 6. Variation in circulation structure size along the back bend: (**a**) Variation in circulation structure size at Q1, (**b**) Variation in circulation structure size at Q2, (**c**) Variation in circulation structure size at Q3. F of the legend denotes the Residual Circulation of the Front Bend (RCFB) in the back bend, and B of the legend denotes the Circulation of the Back Bend (CBB) in the back bend. The grey dotted line is the trend line after removing the out-bank cell. The size of circulation structure here is defined as the area occupied by circulation.

Table 3. Development of the two kinds of circulations in the back bend.

Condition	The Position Corresponding to the Complete Attenuation of the RCFB			The Generation Position of the CBB			The Position at Which the Two Circulation Types Have the Same Intensity		
Discharge	S1	S2	S3	S1	S2	S3	S1	S2	S3
Q1	96°	88°	80°	11.25°	6°	0°	59°	48°	41°
Q2	98°	88°	80°	11.25°	6°	0°	60°	52°	43°
Q3	108°	95°	85°	11.25°	6°	0°	59°	50°	40°

As shown in Figure 6 and Table 3, at Q3~S1, the size of the structure of the RCFB in the back bend decreases gradually along the channel, while the size of the CBB increases. The size of the two kinds of circulation structures has the same intensity at approximately 59°. The trend near the outlet of the back bend has a large variation amplitude, which is mainly influenced by the abovementioned "out-bank cell". The grey dotted line in the figure is the trend line after removing the "out-bank cell". According to its intersection with the abscissa, the position corresponding to the complete attenuation of the RCFB can be deduced. The change in the circulation structure size at other discharges are similar to that of Q3, which also shows that the size of the two kinds of circulation structures increases and decreases along the channel, respectively, but there is no "out-bank cell" near the back bend outlet, and the trend is clearer. The positions where the intensities of the two circulation structures are equal are slightly different at different discharges (see Figure 6a,b).

It can be seen that the size of the two kinds of circulation structures varied with width/depth ratio of the transition section, under different conditions with the same discharge. The size of the RCFB decreases considerably at the inlet section of the back bend and then decreases downstream at the same position, and the complete attenuation position of the RCFB advances gradually. The generation position of the CBB also advances gradually with increasing size. The position where the intensities of the two circulation structures are equivalent clearly advances with decreasing width/depth ratio (see Table 3).

3.2. Analysis of Circulation Intensity

To quantitatively analyse the variation in the circulation on the typical section in the continuous bend, the intensity of the circulation is calculated. In this paper, circulation is regarded as the current rotating around a central point, and vorticity is used as an index to measure the circulation intensity. The vorticity field (three vorticity components) is determined based on the velocity flow field; the calculation formula is as follows [29–32]:

$$\omega_s = \frac{\partial w}{\partial n} - \frac{\partial v}{\partial z} \tag{3}$$

where ω_s is vorticity in downstream direction, v and w are the transverse and vertical components of velocity. After dimensionless treatment, $\omega_s H/U$, the contour map of the circulation intensity distribution in a typical section of a continuous bend was drawn, in which a negative value represents clockwise circulation, a positive value represents anticlockwise circulation, and the 0 contour line represents the boundary line between the two kinds of circulations, which is used to analyse the change in the distribution size of the two kinds of circulations.

Figure 7 shows a three-dimensional view of the distribution of circulation intensity along the continuous bend with Q3. Combined with the analysis of circulation structure, this figure shows that clockwise circulation gradually occurs in the front bend, and the circulation intensity and size increase gradually along the channel, reaching a maximum value near 67.5°; there is no obvious decline in the circulation intensity until the exit of the bend. A small area of positive values appears near the concave bank of the front bend exit, which may indicate the formation of the "out-bank cell", but no circulation structure is formed (see Figure 4). In the transition section, the clockwise circulation is still easily observable, but the intensity, especially the strong circulation intensity, decreases greatly ($\omega_s H/U > 0.6$).

The size of the clockwise circulation at the inlet section of the back bend is still large, almost occupying the whole section. Although the size of the weak circulation intensity decreases to some extent, the size of the strong circulation intensity increases considerably at the 33.75° section, and the overall circulation intensity increases. The strong and weak circulation intensities both decrease rapidly below this section. Although there is sign of anticlockwise circulation at the bottom of the concave pool at the inlet section of the back bend, the formation of the anticlockwise circulation structure does not begin until the section of 11.25° (15+#), after which it develops gradually along the channel, and the intensity and size of the circulation increase gradually. After that, the two anticlockwise circulations merge together, and the circulation intensity centre moves to the middle of the channel.

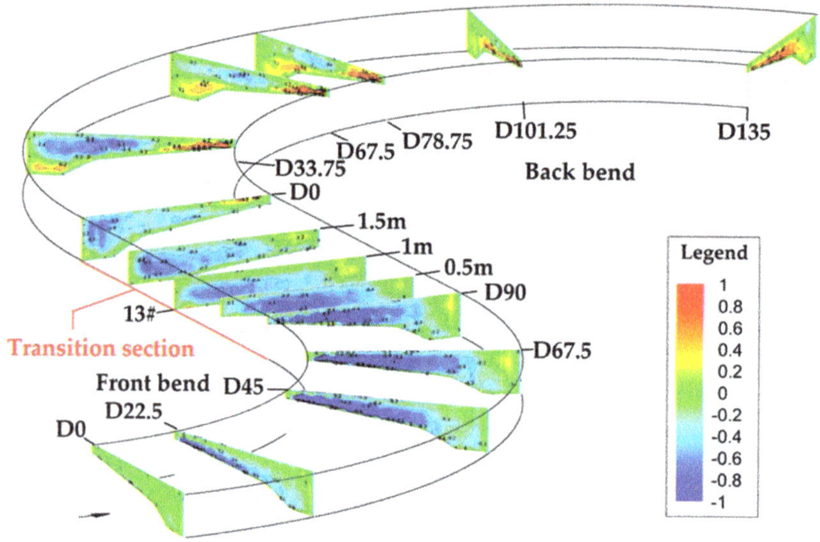

Figure 7. Distribution of the circulation intensity along the continuous bend at Q3.

The sections of the front bend outlet and back bend apex were selected to analyse the variation in circulation intensity at the same time under different scenarios and different discharges. As shown in Figure 8, at Q3 and all three scenarios, the circulation intensity at the outlet section of the front bend weakens slightly, and the left and right sides of the main circulation area reduce slightly, but the overall change is not obvious; however, the circulation intensity and size at the apex section of the back bend change more obviously. With the decrease in the width/depth ratio of the transition section, the clockwise circulation intensity and size gradually decrease, while the counterclockwise circulation intensity and size gradually increase correspondingly. At other discharges, there is an obvious RCFB in the back bend, and the intensity of the CBB is influenced by the front bend; with the decrease in the width/depth ratio of the transition section, the change laws are the same as Q3.

To show the variation in the circulation intensity of the back bend more clearly, circulation intensity distribution of typical cross-sections along the back bend at different scenarios under typical discharge is given in Figure 9. As figure shows, in different scenarios, two kinds of circulation intensity coexist. With the decrease in the width/depth ratio of the transition section, before the cross-section 33.75°, th intensity of the RCFB increases gradually at the same cross-section; then, after this section, all decrease rapidly. A positive intensity circulation appears at the inlet section of the back bend at S3. There are no negative intensity circulations near concave bank water surface at S2 and S3.

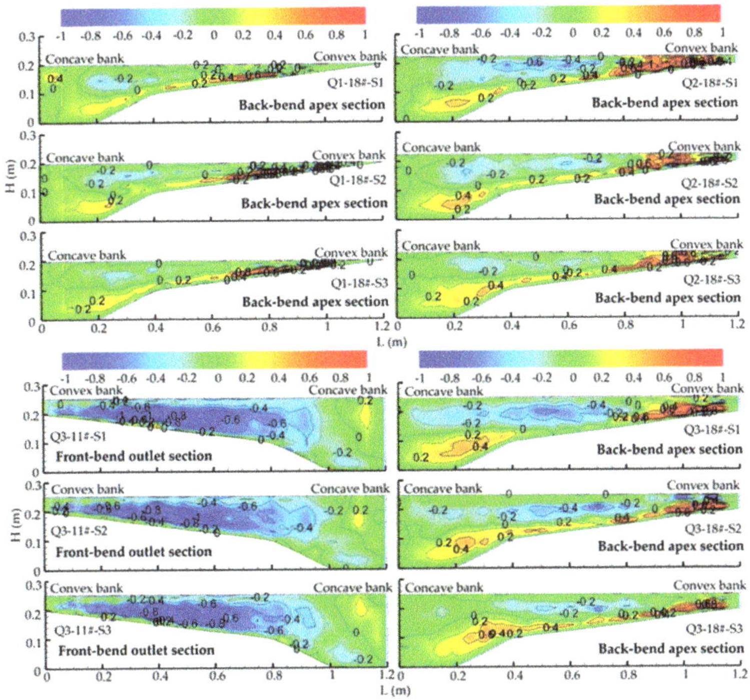

Figure 8. Distribution of the circulation intensity of a typical cross-section under different discharges and scenarios.

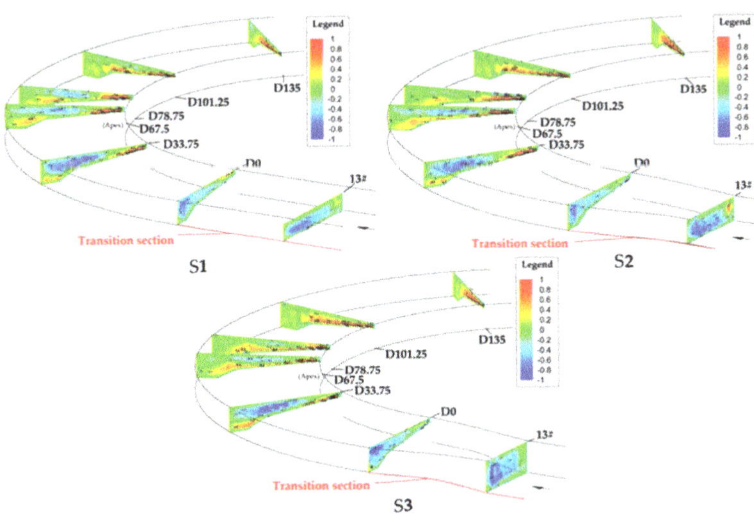

Figure 9. Distribution of the circulation intensity of a typical cross-section along the back bend at Q3 and different scenarios.

Figure 10 shows the variation in the two circulation intensities along the back bend for different discharges and scenarios. At Q3~S1, the intensity of the RCFB increases slightly near the entrance of the back bend and then decreases, with a small amplitude of fluctuation; however, the intensity of the CBB increases abruptly near the entrance and then changes slowly, weakening near the exit of the back bend, with a larger amplitude of fluctuation. The variation in circulation intensity at Q2~S1 is similar to that at Q3, but the variation in the intensity of the RCFB in the forward half of the back bend is not obvious, while at Q1~S1, the intensity of the RCFB is mainly reduced, and the intensity of the CBB mainly increases without fluctuating.

For the same discharge, with the decrease in the width/depth ratio of the transition section, the results of Q2 and Q3 show that the intensity of the RCFB weakens; while the intensity of the CBB changes only slightly, its fluctuation amplitude decreases greatly. In Q1, the intensity of the RCFB changes little, but the intensity of the CBB increases gradually.

At a fixed width/depth ratio, with increasing discharge, the intensity of the RCFB in the back bend gradually increases. The fluctuation of turbulence intensity in the back bend increases, with the distribution becoming more complex.

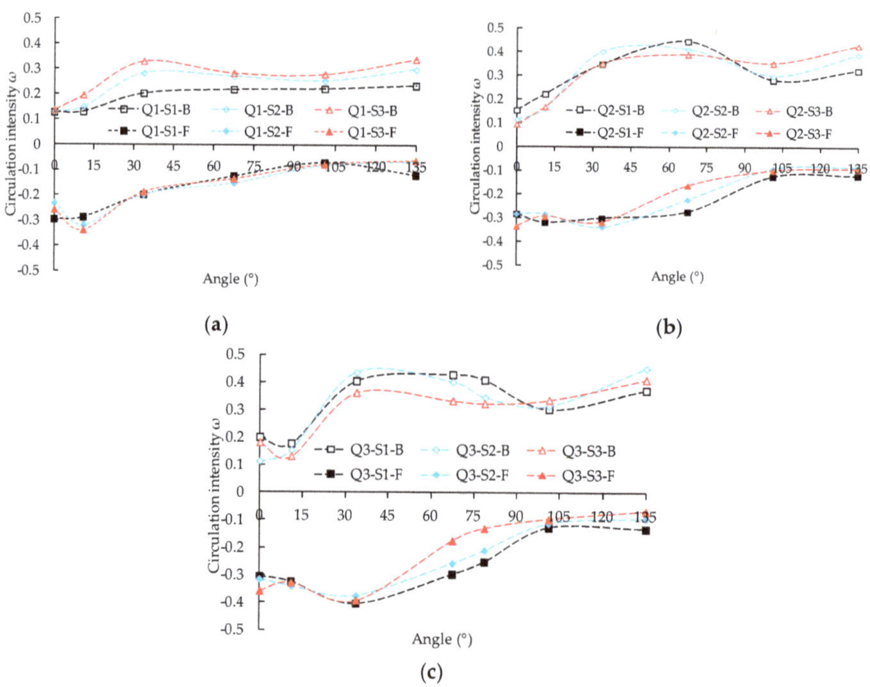

Figure 10. Variation in circulation intensity along the back bend: (**a**) Variation in circulation intensity at Q1, (**b**) Variation in circulation intensity at Q2, (**c**) Variation in circulation intensity at Q3. F of the legend denotes the RCFB in the back bend, and B of the legend denotes the CBB in the back bend.

3.3. Analysis of Turbulent Kinetic Energy

Turbulent kinetic energy is a physical quantity that characterizes the degree of fluid turbulence. The magnitude of turbulent kinetic energy can reflect the intensity of fluid turbulence, the degree of fluid collision and friction. To calculate the TKE of each typical section along the continuous bend, the following calculation formula is employed:

$$k = \frac{1}{2}\left(\overline{u'^2} + \overline{v'^2} + \overline{w'^2}\right) \tag{4}$$

where u', v' and w' are the longitudinal, transverse and vertical turbulence velocity, respectively; $\overline{u'^2}$, $\overline{v'^2}$ and $\overline{w'^2}$ are the longitudinal, transverse and vertical average turbulence intensity, respectively. The TKE distribution diagram of the typical sections along the continuous bend at Q3 is drawn in Figure 11.

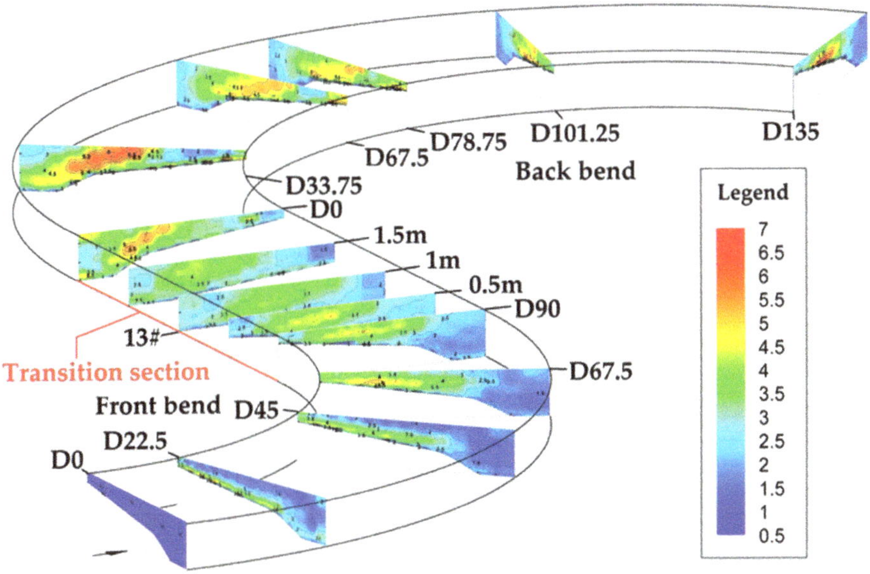

Figure 11. Distribution of Turbulent Kinetic Energy (TKE) along the continuous bend at Q3 (m^2/s^2).

As shown in Figure 11, the TKE in the front bend is divided into two parts, namely, the high-TKE area ($6 > k \geq 4.5$) and the low-TKE area ($k < 4.5$). The results show that the TKE at the inlet section of the front bend is low and then increases gradually along the channel, with the area of high TKE appearing gradually. The high TKE reaches a maximum near 67.5° and then decreases slightly, while the low TKE increases along the channel. In the transition section, the high TKE disappears, and the low TKE increases slightly.

The amplitude of the variation in the distribution of the TKE in the back bend is larger than that in the front bend, with a very high TKE ($k \geq 6$). In the middle part of the flume (0.2–0.9 m), the area of high TKE appears in the upper part of the middle part of the flume at the inlet section, and the overall TKE is relatively high. After that, the TKE in this part of the water body increases greatly, reaching a very high TKE, and then decreases gradually until the areas of very high TKE and high TKE disappear. Correspondingly, the TKE increases gradually in the concave pool; the high-TKE area forms and expands gradually in the lower part of the water body, gradually approaches the middle part of the flume, and remains close to the bed surface. At the same time, with the increase in the TKE, the area of very high TKE slowly forms and then occupies the dominant position. With the area of very high TKE approaching the middle of the flume, the TKE in the concave pool decreases greatly.

As shown in Figure 12, at three scenarios, the high TKE enters the transition section with the flow after exiting the front bend; although the high TKE decreases along the channel, the flow into the back bend still carries a high residual TKE, and after entering the back bend, it experiences rapidly increases. After that, the residual TKE of the front bend begins to be consumed, the back bend begins to produce TKE, which gradually develops along the channel, and the phenomenon of double peaks coexisting in the middle and back reaches of the back bend arises. With the decrease in the width/depth ratio of the transition section, before the cross-section 33.75°, the TKE all increases relative to S1 at the same cross-section; then after this section, the TKE in the concave pool decreases gradually along the way, but the TKE above the convex point bar decreases firstly gradually and then increases relative to S1 at the same cross-section.

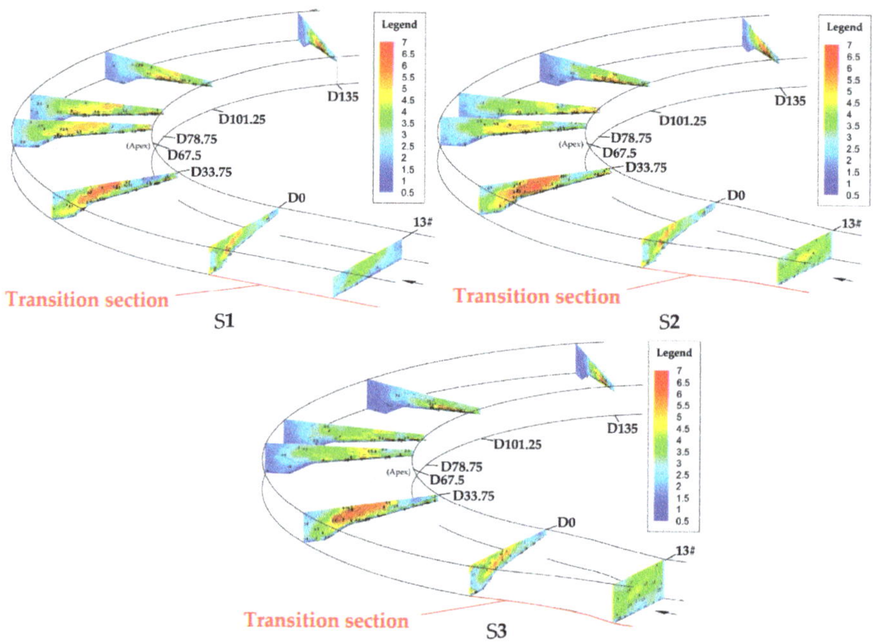

Figure 12. Distribution of the TKE along the back bend at Q3 and at different discharges (m^2/s^2).

Figure 12 shows that the TKE distribution of the back bend is more complex. It is not easy to distinguish the residual TKE of the front bend from that generated by the back bend, and because the TKE is a field concept and a single section is not representative, the total TKE of the whole water body in the back bend is calculated. As shown in Figure 13, according to the change in the total TKE slope along the back bend, at the same discharge, with different width/depth ratios of the transition section, the distribution of the TKE along the back bend water body changes, but the overall change is small. The total TKE shows the same law, increasing slightly at S2 and decreasing slightly at S3, but the overall change is small, with a variation amplitude of approximately 5%. Comparing the results of different discharges under the same conditions, it is found that the greater the discharge is, the greater the amplitude of the variation in the TKE distribution, and the greater the total TKE.

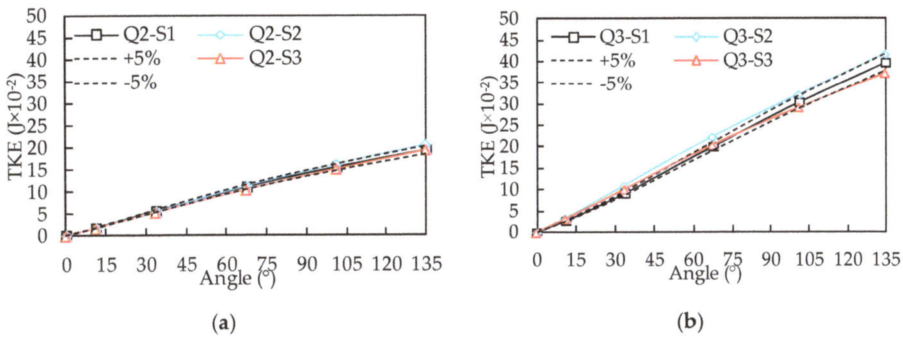

Figure 13. Distribution of TKE in the back bend of the water body: (**a**) Distribution of TKE at Q2, (**b**) Distribution of TKE at Q3. The black dotted line is the 5% variation amplitude line of S1.

3.4. Discussion

In this paper, the correlation between the front and back bends of a continuous bend is mainly revealed by their interaction. However, the analysis of the circulation structure, intensity and distribution of TKE along the continuous bend shows that the influence of the back bend on the front bend is relatively small and not easy to reflect [33], so it is usually not the focus of attention. Moreover, van der Meer et al. [34] and Lanzoni and Seminara [35] have shown that for an open channel with $B/H < 10$, the influence of a change in the downstream riverbed morphology will not extend upstream; the B/H ratios corresponding to all the scenarios in this paper are less than 10. Therefore, this paper mainly discusses the influence of the front bend on the back bend and the correlation reflected from this and explores the influence of the width/depth ratio of the transition section on the correlation between the front and back bends. Here, the correlation characterizes the degree of the influence of the front bend on the back bend, including the influence on circulation structure, circulation intensity and TKE. The greater the degree of influence, the stronger the correlation.

As the results show, when comparing the circulation structures of the front and back bends, there is only one single clockwise circulation in the front bend, and the configuration of the circulation structure is relatively regular. However, in the back bend, two or more circulation structures with different vortex directions coexist, with irregular circulation structure, which is similar to the previous results [22,36]; the circulation structure in the back bend is clearly more complex. The same as the circulation structure, the circulation intensity in the back bend is also clearly more complex. If the back bend is not influenced by the front bend, the evolution law of the circulation intensity of the back bend should be similar to that of the front bend and close to the maximum value near 67.5°. However, the development of the CBB is obviously restrained by the front bend, and the circulation intensity does not reach the maximum value, even at the outlet of the back bend. At the same time, the overall TKE along the back bend is larger, the areas of high TKE and very high TKE are clearly formed, the very high TKE region alternates along the channel, and the TKE distribution in the water body is more complex. The main reason for this phenomenon is the existence of residual TKE carried by the flow into the back bend.

Thus, it can be seen that due to the existence of the front bend and the short length of the transition section, the flow in the back bend carries the RCFB and TKE, which makes the distribution of the circulation structure, intensity and TKE of the back bend more complex, and makes the circulation intensity of the CBB smaller and total TKE greater than that of the front bend (single bend). It can be seen that the presence of the front bend will change the flow characteristics of the back bend. This finding is consistent with the results of Xiao et al. [37] and Moncho Esteve et al. [38]. In addition to the circulation structure and intensity, it can also be seen that there is a strong correlation of the TKE between the front and back bends of the continuous bend.

Comparing different discharges under the same scenarios, it can be seen that with the increase in the discharge, the complete attenuation position of the RCFB gradually moves backwards. The distance of the influence of the front bend on the back bend increases. The intensity of the RCFB in the back bend gradually increases, with the distribution becoming more complex. The greater the amplitude of the variation in the TKE distribution is, the stronger the correlation becomes.

At Q3-S1, the structure of RCFB is continuously consumed when it passes through the transition section and enters the back bend, which is similar to the results of Bryk [25]. Previous studies have indicated that it takes a long distance for the circulation out of the front bend to completely disappear [39,40]. In this study, the complete attenuation position is at approximately 108°. However, according to the decline formula of the out-of-bend circulation presented in previous studies [39], the calculated decay distance is far beyond this length. The main reason for this finding is that the generation of the CBB and the opposite riverbed topography accelerate the decline in the RCFB. It can be seen that the circulations with the opposite vortex directions consume each other in the back bend. However, due to continuous power input, the CBB develops gradually, so the sizes of the two kinds of circulations show decreasing and increasing trends, respectively. The size and intensity of the RCFB decreases gradually, while those of the CBB increases gradually. The intensities of the two kinds of circulations are equivalent at approximately 60°. However, Rui [41] found that the intensities of these two kinds of circulations are equivalent at the outlet of the back bend (90°); the above difference is mainly due to the influence of the transition section, which weakens the front-bend factor before entering the back bend.

As the results shown, the intensity of the RCFB and the TKE all increase firstly (before 33.75°) in the inlet reach of the back bend, then decrease after 33.75°. Termini and Piraino [42] also found that the intensity of the RCFB increases firstly, then it decreases in intensity and disappears at crossover-section downstream, but it increases before the bend apex. However, Abad and Garcia [7,9] found that the intensity of the RCFB keeps increasing.

Combined with the change in circulation structure and intensity, it is found that the change in the TKE is consistent with that of the circulation structure and intensity along the channel, with each high-TKE region corresponding to a region with a strong circulation intensity. There is a single circulation structure in the front bend, whose intensity increases first and then decreases. Same to this study, Shukry [43], Kashyap et al. [44], Bai et al. [45] and Vaghefi et al. [46] all found that the intensity of the circulation increases gradually along the channel, reaching a maximum value at approximately 70°; there is no obvious decline in the circulation intensity until the exit of the bend. Correspondingly, the TKE also exhibits a single peak with similar trends, and even at the same place. The alternation of high-TKE regions in the back bend channel corresponds to the transformation of the primary and secondary positions of the RCFB and CBB. Thus, the size and distribution of the TKE are closely related to the size and distribution of the circulation intensity and circulation structure [47]. Moreover, the structural size, intensity and TKE of the RCFB all shrink to some extent after passing through the transition section, while all of them show expansion trends after entering the back bend. The synchronous trends of the convergence and expansion of structure size, intensity and TKE can also prove the above viewpoint. As shown in the results, with the decrease in the width/depth ratio of the transition section, at all discharges, the size of the RCFB decreases gradually, whereas the size of the CBB increases gradually due to the weakening of the inhibition effect, and the complexity of the distribution of the circulation structure in the back bend decreases; meanwhile, the intensity of the RCFB in the back bend decreases considerably; the complexity of the flow characteristics of the back bend decreases considerably, and the correlation between the front and back bends weakens. By measurement data analysis, You et al. [48] found that barrier reaches with low width/depth ratio can block the upstream channel adjustment from propagating to the downstream reaches in the bifurcated river. You's results indicate that transition sections with low width/depth ratio have barrier effect and can decrease correlation between upstream and downstream reaches, which is in accordance with our results.

However, with the decrease in the width/depth ratio of the transition section, the circulation intensity above the convex point bar of the back bend does not become stronger but weakens gradually. The above phenomena are mainly due to the weakening of the inhibition of the RCFB. The compression effect on the circulation above the convex point bar is weakened accordingly and then extends to the middle of the river channel, so the corresponding vortex rotation degree weakens, which reflects that the circulation intensity decreases. The inhibition of the RCFB in the back bend gradually weakens, the intensity of the CBB also gradually weakens, which is partly due to the change in circulation size, and the overall fluctuation decreases, which indicates that the influence of the front bend gradually weakens. However, Q1 is different from Q2 and Q3: The intensity of the CBB increases gradually, but the variation in the RCFB is small, mainly because the circulation structure does not occupy the whole water body, and the RCFB easily moves; back-circulation develops relatively easily.

With the decrease in the width/depth ratio of the transition section, the intensity of the RCFB that enters the back bend, decreases. The main reason for this is that after decreasing the width/depth ratio of the transition section, the TKE of the water body increases, which leads to an increase in the water mixing level, leading to the energy consumption of the water body. At the same time, a change in the width/depth ratio will cause the river resistance to vary [49,50]. Above a certain width/depth ratio, with decreasing width/depth ratio, the river resistance increases. Under a certain width/depth ratio, although the resistance is not directly affected by the width/depth ratio, the effect of the sidewall becomes larger, and river resistance will also increase [51]. Moreover, in order to reduce the abrupt change in the boundary, the channel shape with a gradual change in river width is adopted; this kind of channel shape will change the flow and turbulence structure, which will directly lead to the direction of flow in a gradually shrinking reach to deviate from the sidewall to the axis and then in a gradually expanding reach to deviate from the axis to the sidewall. The aggregation and diffusion of the streamline increases the mixing level of the water body to some extent, which increases the water energy quantity loss. At the same time, anticlockwise circulation will be produced in the gradually expanding reach [52], which will also increase the consumption of the clockwise circulation of the front bend to some extent.

With the decrease in the width/depth ratio of the transition section, the total TKE in the back bend water body remains almost unchanged. The reason for this may be that when the configuration of the transition section changes, the flow mixing increases, leading to the increasing consumption of the forward bend circulation and increase in the converted TKE, so the TKE near the entrance of the back bend increases. The change in TKE is mainly concentrated in the second half of the transition section [53], and the change in TKE gradually highlights after entering the back bend; therefore, the difference in total TKE entering the back bend for the three scenarios is minor. Only with the decrease in the width/depth ratio of the transition section, part of the energy is carried by the RCFB that is converted into TKE in advance. Moreover, this study shows that the weakening of the RCFB corresponds to the enhancement of the CBB, so the corresponding TKE of the CBB is also slightly increased, which can reduce the difference caused by TKE loss in the transition section observed under all three scenarios. Therefore, the total TKE of the back bend water body does not change significantly with the change in the width/depth ratio of the transition section.

As mentioned above, the correlation is reflected by the influence of the front bend on the back bend, including the influence on the distribution of circulation structure, circulation intensity and TKE, number of the circulation and so on. However, all of these phenomena are due to the direct influence of the RCFB existence; thus, the distance that the RCFB exists in the back bend directly determines the degree of influence, that is, the correlation (see Table 3). The ratio of the existing distance of RCFB in back bend to the length of the back bend is used as a measure of the correlation and as a quantitative indicator of the correlation. The longer the distance of the RCFB in the back bend, that is, the greater the ratio of the existing distance of RCFB in back bend to the length of the back bend, indicates that the greater the influence of the front bend on the back bend, the stronger the correlation between the front and back bends. With the increase of the above ratio, the influence of front bend on back bend increases, and the correlation increases, and vice versa. The variation of the above ratio represents the variation of influence, that is, the variation of correlation between the front and back bends. The results are shown in Table 4: At Q3~S1, the correlation between the front and back bend is 0.8, which shows that the correlation is strong, and with the decrease of the discharges and width/depth ratios of the transition section, the correlations weaken.

Table 4. The correlation between the front bend and back bend and its variation relative to S1 and Q1.

Discharge	Correlation			Variation (Q, %)			Variation (S, %)		
	S1	S2	S3	S1	S2	S3	S1	S2	S3
Q1	0.71	0.65	0.59	\	\	\	\	8.45	16.90
Q2	0.73	0.65	0.59	2.82	0	0	\	11.26	19.18
Q3	0.80	0.70	0.63	12.68	7.69	6.78	\	12.50	21.25

In the experiment, the width/depth ratio of the transition section is reduced twice, and the variation of the correlation between the front and back bend is analysed and discussed (Table 4). In the first decrease in the width/depth ratio, the variation of the correlation is relatively large, 1~1.43 times that in the second one, and on the whole, the differences (including the position of the same intensity of the two kinds of circulations (Table 3)) are not significant. However, the corresponding width/depth ratio reduces to 33% of the original channel for the first decrease and only 9.5% for the second decrease. It can be seen that the relationship of the decrease in the width/depth ratio of the transition section and the decrease in the correlation of the front and back bends is not linear but approximately exponential. That is, the smaller the width/depth ratio is, the greater the influence of the same amplitude of variation in the width/depth ratio.

The relationship between the influence of the width/depth ratio on the correlation between the front and back bends of continuous bends and the discharge is discussed and analysed. It is found that there is also no linear relationship between the amplitude of the variation in discharge and the correlation; the larger the discharge is, the smaller the increase in the amplitude of the variation in the correlation caused by the same amplitude of the variation in the discharge. Moreover, with the same width/depth ratio, the greater the discharge is, the greater the influence of the width/depth ratio of the transition section on the correlation between the front and back bends. The relationship between the amplitude of the variation in the width/depth ratio and the correlation, as well as the relationship between the amplitude of the variation in the discharge and the correlation, are explored; the former shows an approximate exponential relationship, while the latter is close to a linear relationship. The influence of the width/depth ratio on the correlation is clearly greater than that of the discharge.

The flow characteristics of any single bend in a continuous bend are influenced not only by its own characteristics (configuration and composition of riverbed boundary) but also by the upstream and downstream reaches. In this paper, the relationship between the front and back bends of continuous bends and the influence of transition sections are investigated. We found that the upstream reach has a significant influence on the downstream reach, which leads to the obvious differences in the flow characteristics of the downstream reach in the continuous bend. The decrease of the width/depth ratio of the transition section will weaken the correlation between the front and back bends of the continuous bend. Hooke [54] observed that there was no correlation between continuous curves due to obstacles or contraction points between curves, which aligns with our findings. The different flow characteristics caused by the presence of front bend can inevitably lead to the differences of various forces in the water body and then affect the channel evolutions. That is, the correlation of flow movement characteristics between the front and back bends of continuous bends is bound to correspond to the correlation of river bed evolution, which is similar to the conclusion obtained by Frothingham and Rhoads [2]. By contrast, Lotsari et al. [55] found that there is neither correlation of flow characteristics nor the riverbed evolution between the front and back bends of a continuous bend, and thus thought that the single bend and its upstream and downstream reaches in a continuous bend are independent of each other in spatial evolution. Given the current different viewpoints on this [2,55], therefore, we suggest to analyse the bed shear stress and conduct movable bed experiments to further investigate the correlation between the front and back bends of continuous bends.

4. Conclusions

In order to better understand the influence of transition section configuration on the correlation of flow movement characteristics between the front and back bends of a continuous bend, laboratory flume experiments were implemented. The three-dimensional instantaneous velocity was measured by Acoustic Doppler Velocimeter (ADV). Through the analysis of circulation structure, circulation intensity and turbulent kinetic energy, the correlation of flow movement characteristics between the front and back bends in a continuous bend under different width/depth ratios of transition section were studied. The main findings of this study are the following:

1. In addition to those of the circulation structure and circulation intensity, there is a strong correlation of the TKE between the front and back bends of the continuous bend, which shows that the TKE of the back bend water body is higher as a whole than that of the front bend (single bend), and the distribution of the TKE across each section along the channel is more complex.
2. When the width/depth ratio is constant and the discharge increases, the influence level of the RCFB increases, the influence distance becomes longer, and the correlation of the flow movement between the front and back bends is enhanced. The larger the discharge, the greater the amplitude of the variation in the correlation caused by the same discharge variation amplitude. Moreover, the greater the discharge is, the greater the influence of the same amplitude of variation in the width/depth ratio of the transition section on the correlation between the front and back bends.
3. At the same discharge, with the decrease in the width/depth ratio, the size and intensity of the residual circulation at the same position in the back bend decrease; the decay of the residual circulation in the back bend accelerates, and the position corresponding to complete attenuation advances; the corresponding position of CBB generation advances, and its size and intensity increase; the distribution of TKE along the back bend water body adjusts, but the total TKE remains unchanged. On the whole, the decrease in the width/depth ratio leads to the weakening of the influence of the front bend on the back bend and the correlation between the front and back bends. Moreover, the smaller the width/depth ratio is, the greater the influence of the same amplitude of variation in the width/depth on the correlation. Within the scope of this paper, the most favourable transition geometrical configuration was the one with the minimum average width/depth ratio, the width/depth ratio decreases by 42.5% with the correlation decreased by approximately 20%.
4. There is no linear relationship between the amplitude of the variation in the discharge and the correlation between the front and back bends, but when the discharge increases, the correlation corresponding to the same variation in the discharge amplitude strengthens slightly. The relationship between the decrease in the amplitude of the width/depth ratio and the correlation is also not linear; it is approximately exponential. This comparison shows that the influence of the width/depth ratio on the correlation is greater than that of the discharge.

Overall, our study reveals the important role of transition sections in a continuous bend and may shed light on the mechanics of flow in continuous bends. It can also provide theoretical support for further research on sediment transport and riverbed evolution of continuous bends.

Author Contributions: Conceptualization, K.Z. and L.C.; data curation, Y.L., B.Y. and Y.W.; formal analysis, K.Z.; funding acquisition, L.C.; investigation, K.Z.; methodology, K.Z. and Y.L.; project administration, L.C.; resources, L.C.; software, K.Z., Y.L. and B.Y.; visualization, K.Z.; writing—original draft, K.Z.; writing—review and editing, L.C. All authors have read and agreed to the published version of the manuscript.

Funding: This research was funded by the National Key Research and Development Program of China (Grant No. 2016YFC0402101).

Conflicts of Interest: The authors declare no conflict of interest.

Nomenclature

TKE	Turbulent Kinetic Energy
RCFB	Residual Circulation of the Front Bend
r	Radius of the bend
L/R	Relative transition section length
ADV	Acoustic Doppler Velocimeter
R/B	Curvature ratio
U	Average velocity of the cross-section
H	Average depth of the cross-section
g	Gravitational acceleration
R	Hydraulic radius
v	Kinematic viscosity coefficient
S	Scenario
Q	Discharge
h	Elevation
Fr	Froude number
$\overline{R}$	Radius of the central line
B	Channel width of the continuous bend
B^*	Width of section 13#
H^*	Flow depth of section 13#
R^*	Hydraulic radius of section 13#
U^*	Average velocity of section 13#
B^*/H^*	Width/depth ratio of section 13#
$\overline{B}$	Mean channel width of the transition section
$\overline{H}$	Mean channel depth of the transition section
$\overline{B}/\overline{H}$	Mean width/depth ratio of the transition section
Re	Reynolds number
s	Longitudinal coordinates
n	Transverse coordinates
z	Vertical coordinates
SNR	Signal-to-Noise Ratio
h^*	Water depth
CBB	Back Bend Circulation
v	Transverse component of velocity
w	Vertical component of velocity
w_s	Vorticity
w_sH/U	Dimensionless vorticity
u'	Longitudinal turbulence velocity
v'	Transverse turbulence velocity
w'	Vertical turbulence velocity
$\overline{u'^2}$	Longitudinal average turbulence intensity
$\overline{v'^2}$	Transverse average turbulence intensity
$\overline{w'^2}$	Vertical average turbulence intensity

References

1. Rowley, T.; Konsoer, K.M.; Ursic, M.; Langendoen, E.J. Evaluating interactions between flow structure and planform evolution on a series of six consecutive meander bends along the Pearl River, Louisiana. *AGUFM* **2018**, *2018*, EP33D-2442.

2. Frothingham, K.M.; Rhoads, B.L. Three-dimensional flow structure and channel change in an asymmetrical compound meander loop, Embarras River, Illinois. *Earth Surf. Process. Landf.* **2003**, *28*, 625–644. [CrossRef]
3. Mockmore, C. Flow around Bends in Stable Channels. *Trans. Am. Soc. Civ. Eng.* **1943**, *3*, 334.
4. Huang, S.L.; Jia, Y.F.; Chan, H.C.; Wang, S.S.Y. Three-Dimensional Numerical Modeling of Secondary Flows in a Wide Curved Channel. *J. Hydrodyn. Ser. B* **2009**, *21*, 758–766. [CrossRef]
5. Stoesser, T.; Ruether, N.; Olsen, N.R.B. Calculation of primary and secondary flow and boundary shear stresses in a meandering channel. *Adv. Water Resour.* **2010**, *33*, 158–170. [CrossRef]
6. Baek, K.O.; Seo, I.W.; Jeong, S.J. Evaluation of Dispersion Coefficients in Meandering Channels from Transient Tracer Tests. *J. Hydraul. Eng.* **2006**, *132*, 1021–1032. [CrossRef]
7. Abad, J.D.; Garcia, M.H. Experiments in a high-amplitude Kinoshita meandering channel: 2. Implications of bend orientation on bed morphodynamics. *Water Resour. Res.* **2009**, *45*, 142–143. [CrossRef]
8. Blanckaert, K.; Kleinhans, M.G.; McLelland, S.J.; Uijttewaal, W.S.J.; Murphy, B.J.; van de Kruijs, A.; Parsons, D.R.; Chen, Q. Flow separation at the inner (convex) and outer (concave) banks of constant-width and widening open-channel bends. *Earth Surf. Process. Landf.* **2013**, *38*, 696–716. [CrossRef]
9. Abad, J.D.; Garcia, M.H. Experiments in a high-amplitude Kinoshita meandering channel: 1. Implications of bend orientation on mean and turbulent flow structure. *Water Resour. Res.* **2009**, *45*, 7–25. [CrossRef]
10. Kang, S.; Sotiropoulos, F. Flow phenomena and mechanisms in a field-scale experimental meandering channel with a pool-riffle sequence: Insights gained via numerical simulation. *J. Geophys. Res. Earth Surf.* **2011**, *116*. [CrossRef]
11. Ghanmi, A. Modeling of flows between two consecutive reverse curves. *J. Hydraul. Res.* **1999**, *37*, 121–135. [CrossRef]
12. Liu, Y.Q. Turbulence Intensity of Bend Flow. *J. South China Univ. Technol. Nat. Sci.* **2003**, *12*, 89–93. (In Chinese)
13. Hu, X.Y.; Zhang, Q.S.; Ma, L.J. Numerical simulation of influence of transition section on continuously curved channel flow. *Adv. Water Sci.* **2011**, *22*, 851–858. (In Chinese)
14. Zhou, J.Y.; Shao, X.J.; Wang, H.; Jia, D.D. Assessment of the predictive capability of RANS models in simulating meandering open channel flows. *J. Hydrodyn. Ser. B* **2017**, *29*, 40–51. [CrossRef]
15. Zhou, J.Y.; Shao, X.J.; Yang, Y.; Qin, C.C.; Jia, D.D. Assessing the Predictive Capabilities of Several RANS Turbulent Models for Flow in Consecutive Reverse Bends. In *Proceedings of the 35th Iahr World Congress, Vols I and Ii. Int. Assoc. Hydro Environ. Eng. & Res., Chengdu, China, 8–13 September 2013*; Zhaoyin, W., Lee, J.H.W., Eds.; Tsinghua University: Beijing, China, 2013.
16. Jing, X.J.; Zhao, Z.Z. Experimental Study on the Effect of Cross-Sectional Shape of Transition Section on the Transverse Slope in Successive Bends. *J. Chongqing Jiaotong Univ.* **2013**, *32*, 681–686. (In Chinese)
17. Cai, J.D.; Wang, W.; Wu, X.L. A Study on Bed-load Exchange between the Upper Pool and the Crossing Bar in Curved River. *J. Sediment. Res.* **1993**, *18*, 95–102. (In Chinese)
18. Wang, X.Y.; Li, Y.T.; Sun, Z.H. Study on the regulation width of the middle and lower reaches of the Yangtze River. *J. Sediment. Res.* **2005**, *30*, 36–41. (In Chinese)
19. Yin, Y.Q.; Hu, X.Y. Analysis on navigation parameters and straightening distances of ships passing through continuous bends. *J. Transp. Sci. Eng.* **2019**, *35*, 85–90. (In Chinese)
20. Dietrich, W.E.; Smith, J.D. Influence of the point bar on flow through curved channels. *Water Resour. Res.* **1983**, *19*, 1173–1192. [CrossRef]
21. Nelson, J.M.; Smith, J.D. Evolution and Stability of Erodible Channel Beds. In *River Meandering*; Ikeda, S., Parker, G., Eds.; American Geophysical Union: Washington, DC, USA, 1989; Volume 12, pp. 321–377.
22. Thorne, C.R.; Hey, R.D. Direct measurements of secondary currents at a river inflexion point. *Nature* **1979**, *280*, 226–228. [CrossRef]
23. Wang, J.Z.; Chen, L.; Zhang, W.; Chen, F. Experimental study of point bar erosion on a sand-bed sharp bend under sediment deficit conditions. *Sediment. Geol.* **2019**, *385*, 15–25. [CrossRef]
24. Blanckaert, K. Hydrodynamic processes in sharp meander bends and their morphological implications. *J. Geophys. Res. Earth Surf.* **2011**, *116*, F01003. [CrossRef]
25. Bryk, A.B. The Influence of Channel-Skewed Bedforms on Flow Structure in a High-Amplitude Meandering Channel. Ph.D. Thesis, University of Illinois, Urbana-Champaign, IL, USA, 2018.
26. Nikora, V.I.; Goring, D.G. ADV Measurements of Turbulence: Can We Improve Their Interpretation? *J. Hydraul. Eng.* **1998**, *124*, 630–634. [CrossRef]
27. Goring, D.G.; Nikora, V.I. Despiking Acoustic Doppler Velocimeter Data. *J. Hydraul. Eng.* **2002**, *128*, 117–126. [CrossRef]

28. Esfahani, F.S.; Keshavarzi, A. Circulation cells topology and their effect on migration pattern of different multi-bend meandering rivers. *Int. J. Sediment. Res.* **2020**, *35*, 636–650. [CrossRef]
29. Duarte, A. An Experimental Study on Main Flow, Secondary Flow and Turbulence in Open-Channel Bends with Emphasis on Their Interaction with the Outer-Bank Geometry. Ph.D. Thesis, Ecole Polytech. Fed. Lausanne, Lausanne, Switzerland, 2008.
30. De Vriend, H.J. Steady Flow in Shallow Channel Bends. Ph.D. Thesis, Delft University of Technology, Delft, The Netherlands, 1981.
31. De Vriend, H.J. Velocity redistribution in curved rectangular channels. *J. Fluid Mech.* **1981**, *107*, 423. [CrossRef]
32. Blanckaert, K.; de Vriend, H.J. Secondary flow in sharp open-channel bends. *J. Fluid Mech.* **2004**, *498*, 353–380. [CrossRef]
33. Rodriguez, J.F.; Bombardelli, F.A.; García, M.H.; Frothingham, K.M.; Rhoads, B.L.; Abad, J.D. High-resolution numerical simulation of flow through a highly sinuous river reach. *Water Resour. Manag.* **2004**, *18*, 177–199. [CrossRef]
34. Van der Meer, C.; Mosselman, E.; Sloff, C.; Jagers, B.; Zolezzi, G.; Tubino, M. Numerical simulations of upstream and downstream overdeepening. In Proceedings of the 7th Symposium of River, Coastal and Estuarine Morphodynamics, Beijing, China, 6–8 September 2011.
35. Lanzoni, S.; Seminara, G. On the nature of meander instability. *J. Geophys. Res. Earth Surf.* **2006**, *111*, F04006. [CrossRef]
36. Chacinski, T.M.; Francis, J.R.D.; Werner, P.W. Discussion of "on the origin of River Meanders". *Trans. Am. Geophys. Union* **1952**, *33*, 771. [CrossRef]
37. Xiao, Y.; Yang, F.S.; Fu, H.X.; Li, J.W. Performance of Three-Dimensional Modeling for Flow Structures in Channel Bends. *J. Waterw. Port Coast. Ocean Eng.* **2017**, *143*, 06017004. [CrossRef]
38. Moncho-Esteve, I.J.; García-Villalba, M.; Muto, Y.; Shiono, K.; Palau-Salvador, G. A numerical study of the complex flow structure in a compound meandering channel. *Adv. Water Resour.* **2018**, *116*, 95–116. [CrossRef]
39. Nouh, M.; Townsend, R.D. Shear-stress distribution in stable channel bends. In *Journal of the Hydraulics Division*; John, A.H., Melvin, W.A., Eds.; American Society of Civil Engineers: New York, NY, USA, 1979; Volume 105, pp. 1233–1245.
40. Li, Q.; Song, J.; Li, C.; Wei, Y.; Chen, J. Numerical and experimental study of particle deposition on inner wall of 180° bend. *Powder Technol.* **2013**, *237*, 241–254. [CrossRef]
41. Rui, D.F. *Numerical Model for Circumfluence and Movement of Sediment in Continuous Meandering River*; Sichuan Univ.: Sichuan, China, 2005. (In Chinese)
42. Termini, D.; Piraino, M. Experimental analysis of cross-sectional flow motion in a large amplitude meandering bend. *Earth Surf. Process. Landf.*. **2011**, *36*, 244–256. [CrossRef]
43. Shukry, A. Flow around bends in an open flume. *Trans. Am. Soc. Civ. Eng.* **1950**, *115*, 751–788.
44. Kashyap, S.; Constantinescu, G.; Rennie, C.D.; Post, G.; Townsend, R. Influence of Channel Aspect Ratio and Curvature on Flow, Secondary Circulation, and Bed Shear Stress in a Rectangular Channel Bend. *J. Hydraul. Eng.* **2012**, *138*, 1045–1059. [CrossRef]
45. Bai, Y.C.; Song, X.L.; Gao, S.X. Efficient investigation on fully developed flow in a mildly curved 180° open-channel. *J. Hydroinf.* **2014**, *16*, 1250–1264. [CrossRef]
46. Vaghefi, M.; Akbari, M.; Fiouz, A.R. An experimental study of mean and turbulent flow in a 180 degree sharp open channel bend: Secondary flow and bed shear stress. *KSCE J. Civ. Eng.* **2015**, *20*, 1582–1593. [CrossRef]
47. Zeng, J.; Constantinescu, G.; Blanckaert, K.; Weber, L. Flow and bathymetry in sharp open-channel bends: Experiments and predictions. *Water Resour. Res.* **2008**, *44*, W09401. [CrossRef]
48. You, X.; Tang, J.; Zhang, X.; Hou, W.; Yang, Y.; Sun, Z.; Weng, Z. The mechanism of barrier river reaches in the middle and lower Yangtze River. *J. Geogr. Sci.* **2017**, *27*, 1249–1267. [CrossRef]
49. Knight, D.W. Boundary shear in smooth and rough channels. *J. Hydraul. Div.* **1981**, *107*, 839–851.
50. Knight, D.W.; Demetriou, J.D.; Hamed, M.E. Boundary shear in smooth rectangular channels. *J. Hydraul. Eng.* **1984**, *110*, 405–422. [CrossRef]
51. Church, M.; Biron, P.M.; Roy, A.G. Gravel Bed Rivers: Processes, Tools, Environments. *Aust. Geogr.* **2012**, *45*, 565–567.
52. Wang, S.Y.; Zhou, S.F.; Zhao, X.E.; Liu, X.N.; Wang, X.K. Experimental study on the flow characteristics at local diverging-converging sections in mountain lotus root shape channel. *J. Sichuan Univ.* **2013**, *45*, 51–54. (In Chinese)

53. Wang, W.E.; Liao, W.; Qi, L.J. Experimental study on the turbulence characteristics of wide and narrow river channels. *Adv. Water Sci.* **2020**, *31*, 394–403. (In Chinese)
54. Hooke, J. Spatial variability, mechanisms and propagation of change in an active meandering river. *Geomorphology* **2007**, *84*, 277–296. [CrossRef]
55. Lotsari, E.; Vaaja, M.; Flener, C.; Kaartinen, H.; Kukko, A.; Kasvi, E.; Hyyppä, H.; Hyyppä, J.; Alho, P. Annual bank and point bar morphodynamics of a meandering river determined by high-accuracy multitemporal laser scanning and flow data. *Water Resour. Res.* **2014**, *50*, 5532–5559. [CrossRef]

water

MDPI

Article

Comparison of Methods for Bed Shear Stress Estimation in Complex Flow Field of Bend

Liyuan Zhang, Faxing Zhang *, Ailing Cai, Zhaoming Song and Shilin Tong

State Key Laboratory of Hydraulics and Mountain River Engineering, Sichuan University, Chengdu 610065, China; Zhangliyuan_SCU@163.com (L.Z.); caidouhua_scu@163.com (A.C.); szming970116@163.com (Z.S.); tsl_1996@163.com (S.T.)

* Correspondence: zhfx@scu.edu.cn

Received: 16 August 2020; Accepted: 30 September 2020; Published: 2 October 2020

Abstract: Bed shear stress is closely related to sediment transport in rivers. Bed shear stress estimation is very difficult, especially for complex flow fields. In this study, complex flow field measurement experiments in a 60° bend with a groyne were performed. The feasibility and reliability of bed shear stress estimations using the log-law method in a complex flow field were analyzed and compared with those associated with the Reynolds, Turbulent Kinetic Energy (TKE), and TKE-w′ methods. The results show that the TKE, Reynolds, and log-law methods produced similar bed shear stress estimates, while the TKE-w′ method produced larger estimates than the other methods. The TKE-w′ method was found to be more suitable for bed shear stress estimation than the TKE method, but the value of its constant C_2 needed to be re-estimated. In a complex, strong, three-dimensional flow field, the height of the measurement point (relative or absolute) should be re-estimated when a single point measurement is used to estimate the bed shear stress. The results of this study provide guidance for experimental measurement of bed shear stress in a complex flow field.

Keywords: 60° bend; bed shear stress; turbulent kinetic energy; Reynolds shear stress; log-law

1. Introduction

Bed shear stress is a basic variable in the study of fluvial processes [1]. It is very difficult to estimate bed shear stress via experimental methods, particularly for complex, three-dimensional flow fields. The resistance balance method is the basic method used for calculating bed shear stress, but it is not suitable for local shear stress estimations. In addition, for a flume with limited length, measurements of the water surface profile often have large errors, which can greatly influence the calculated value of bed shear stress. The local mean bed shear stress can also be measured by a shear plate sensor in small-scale and large-scale laboratory flumes [2].

With the development of advanced velocity measurement devices such as the Acoustic Doppler Velocimeter (ADV), many researchers have calculated bed shear stress by measuring flow velocity directly. The local bed shear stress can be calculated by the Turbulent Kinetic Energy (TKE) method [3] as follows:

$$\tau_0 = \frac{1}{2}C_1\rho(\overline{u'^2} + \overline{v'^2} + \overline{w'^2}) \tag{1}$$

where u', v', and w' are the velocity fluctuations of the streamwise, transverse, and vertical components ($u' = u - \overline{u}$, $v' = v - \overline{v}$, and $w' = w - \overline{w}$), respectively, and C_1 is a proportionality constant (0.19).

Kim et al. [3] proposed a modified Turbulent Kinetic Energy method (TKE-w′ method) that uses only vertical velocity fluctuations, as ADV noise errors associated with vertical velocity variances are smaller than the noise errors associated with horizontal velocity fluctuations (Voulgaris and Trowbridge, [4]).

$$\tau_0 = C_2\rho(\overline{w'^2}) \tag{2}$$

where C_2 is a proportionality constant (0.9).

The bed shear stress values are generally considered close to the Reynolds stress values near the bed. The local bed shear stress can also be calculated by the Reynolds stress (Reynolds method) [3,5]:

$$\tau_0 = -\rho(\overline{u'w'}) \tag{3}$$

Where possible, the bed shear stress is usually extrapolated from the Reynolds stress at the bed [6,7]. Reynolds stress values measured directly near the bed can also be used [8].

For a flow with a bed, when the mainstream has a fully developed turbulence flow, the flow can be divided into an inner region and an outer layer, depending on the normal distance from the bed (Figure 1). The outer layer is the fully developed turbulence flow. In the inner region, the flow is different from that in the outer layer because of the influence of the bed. The inner region can be divided into three sub-layers. Theoretically, the bed shear stress can be determined from measuring the velocity of the viscous sub-layer, according to the linear distribution of the velocity. However, because of the very thin viscous sub-layer, the normal distance from the bed is usually on the order of 0.1 mm, which is difficult to measure. PIV has been used to measure the velocity of viscous sub-layers, but the velocity of the flow should be limited to a low-velocity range. In addition, PIV has strict requirements for the measuring environment. Light conditions, sampling frequency, sampling size, etc., would affect the measurement results, and even the temperature of the water must be controlled within a stable range. Otherwise, changes to water temperature may also cause displacement of the laser disc and camera focus plane [9]. The viscous shear stress and turbulent shear stress in the buffer region are equivalent, the flow structure is complicated, and the velocity distribution cannot be expressed by a formula or law. In the turbulent log-law region, turbulent shear stress dominates, and the velocity distribution in this region conforms to logarithmic law distribution. Because the viscous sub-layer and buffer region account for a small proportion of the total flow depth, the velocity distribution of the full flow depth in a simple open channel generally conforms to a logarithmic law distribution; thus, bed shear stress can be obtained by fitting the logarithmic law to the flow velocity (the log-law method) [1,10]:

$$\frac{u}{u_*} = \frac{1}{\kappa}\ln\left(\frac{z_a}{z_0}\right) \tag{4}$$

where u is the velocity, u_* is the friction velocity ($= \sqrt{\frac{\tau_0}{\rho}}$), z_a is the height above the bed, κ is von Karman's constant (0.4), and z_0 is the statistical roughness length.

Application of this method only requires measuring the time-averaged velocity to obtain the friction velocity u_*, using a series of measured (u, z_a) values to carry out logarithmic fitting, which greatly reduces the influence of larger individual error velocity points. Although this method is widely used, there remains a problem with the accuracy of data fitting [11]. In a complex flow field, the velocity distribution of the total depth typically does not satisfy the logarithmic law distribution. The number of measured velocity points and the height error of the measuring points greatly influence the calculation results.

There have been many studies on bed shear stress in the case of uniform flow [12,13], but few studies have estimated bed shear stress in cases of complex flow fields, especially complex flow fields in bends [14]. However, natural rivers, in practice, often have bends, and the flow pattern is complex [15]. In a complex flow field, although it is difficult to satisfy the logarithmic distribution of the total flow depth, the bed shear stress can also be calculated using the velocity distribution of the turbulent log-law region near the bed. In this study, based on the flow field measurements of a 60° bend movable bed experiment, the feasibility and reliability of the log-law method in the complex flow field of the bend were analyzed, and the results were compared with those predicted by the Reynolds, TKE, and TKE-w′ methods.

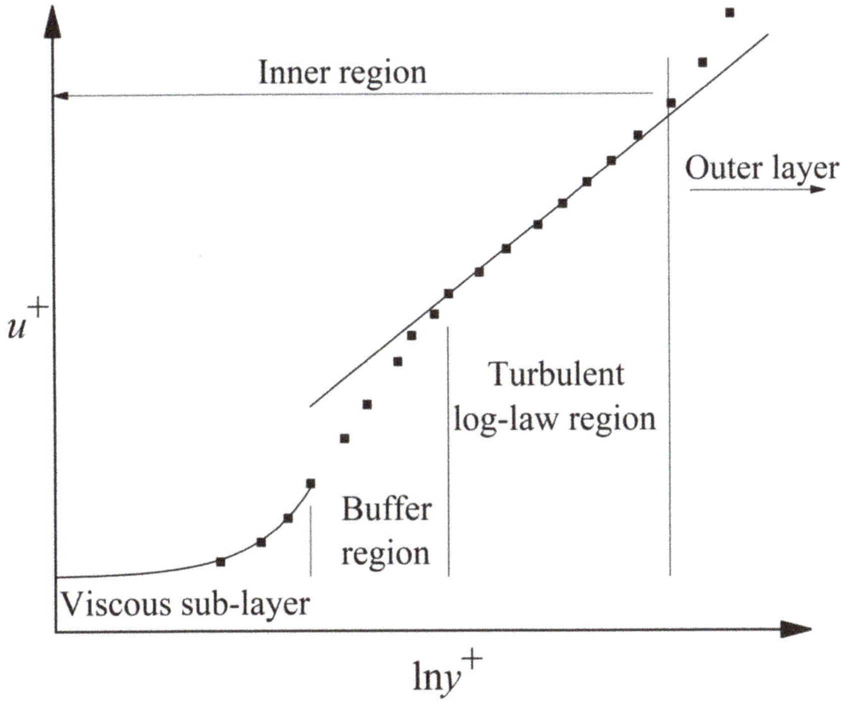

Figure 1. Velocity distribution near the bed ($u^+ = u/u_*$, $y^+ = \frac{u_* y}{\nu}$).

2. Experimental Setup

The experiment was carried out in a 0.6 m wide, 14.4 m long recirculating flume with a longitudinal slope (i) of 0.001, located in the Sichuan University's State Key Laboratory of Hydraulics and Mountain River Engineering. The main channel consisted of a 7 m long upstream, a 60° channel bend with a centerline radius bend (R_c) of 4.2 m, and a 3 m long downstream straight reach. The groyne, with a thickness of 0.01 m, was positioned at the 30° center corner of the outer bank side of the bend. The length of the groyne wing and web was 0.1 m (Figure 2a). To stabilize the flow, an energy dissipation grid was placed at the inlet of the flume. The measured cross-sections $S0$–$S33$ corresponded to the central angles of 0–33°, respectively, of the bend. Figure 2c shows a vertical view of the section measurements in section $S29.5$, and Figure 2d is a three-dimensional illustration of the groyne with a submergence ratio $(H - h)/h = 1$.

Table 1 lists the experimental flow conditions and values of some experimental parameters, such as the median sediment particle size d_{50}. Clear water scouring was used in the experiment. The ratio of the velocity of the approaching flow U_0 to the critical velocity of the sediment v_c was 92.5%. Because the sediment was sand, the critical velocity of the sediment was calculated by the Zhang Ruijin formula as defined in Equation (5) [16]:

$$v_c = \left(\frac{H}{d_{50}}\right)^{0.14} \sqrt{17.6\frac{\gamma_s - \gamma}{\gamma} d_{50} + 0.000000605\frac{10 + H}{d_{50}^{0.72}}} \quad (5)$$

where v_c is the critical velocity of the sediment (m/s), d_{50} is the median sediment particle size (m), H is the approaching flow depth (m), and γ_s and γ are the unit weights of the sand and water, respectively (kN/m^3).

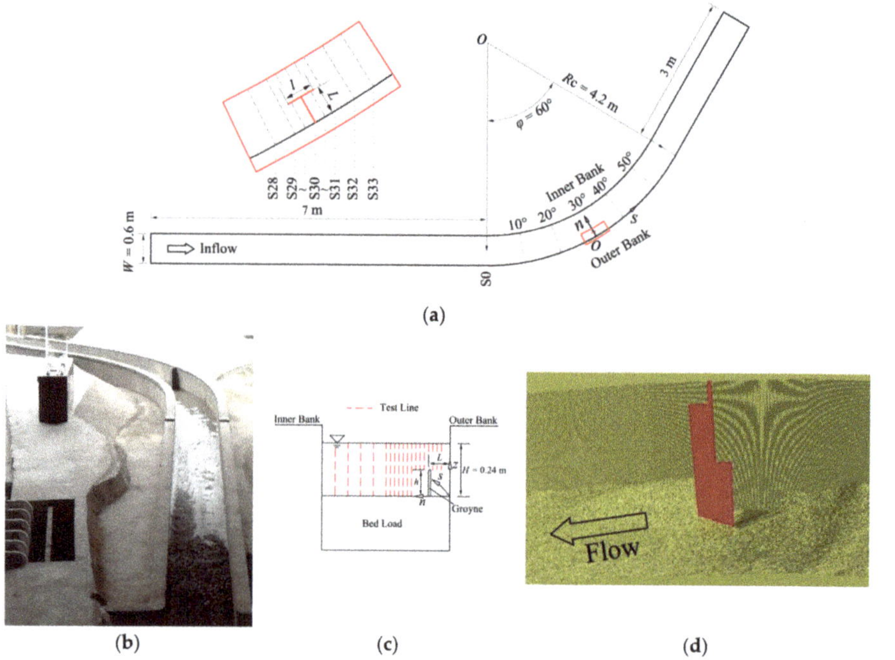

Figure 2. Details of experiment setup: (**a**) plan view parameters related to the experimental flume and arrangement of the cross-sections; (**b**) experimental flume; (**c**) test lines (*S*29.5); (**d**) groyne.

Table 1. Experimental parameter values.

Q (m^3/s)	Fr	H (m)	Rc/W	d_{50} (m)	γ_s (kN/m^3)	γ (kN/m^3)	U_0 (m/s)	v_c (m/s)	U_0/v_c
0.0428	0.19	0.24	7	0.000642	25.8	9.8	0.297	0.321	92.5%

Because $U_0/v_c < 1$, the bed sediment is basically stable in the upstream section, while in the bend of the groyne, the velocity increases because of the reduction in the flow cross-section, thus achieving clean water scouring. The experiment was designed to ensure that the sediment in the test sections reached an equilibrium state during the test period of 86 h.

The topographic measurement instrument used was a RIEGL VZ400 laser scanner with a maximum scanning distance of 600 m and a single-point scanning accuracy of 2 mm at 100 m. The instrument had good applicability in measuring moving bed experiments [17]. A Vectrino Profiler manufactured by Nortek Corporation was used for velocity measurements. This device consists of a down-looking measurement probe with a non-measurable region of 4 cm, a signal regulator, and a signal processor. The device can measure velocity within a range of 3 cm at one time, with a minimum layer thickness of 1 mm (that is to say, the velocity of 30 measuring points can be measured at the same time), an accuracy of ±1 mm/s, and a sampling frequency of 100 Hz. In the experiment, 3000 velocity samples were collected at one time. In other studies, some researchers have removed measurements with correlation coefficients below 70% and signal-to-noise ratios (SNRs) below 15 dB when processing data [13,14]. Considering that the validity and accuracy of the bed shear stress calculation in this experiment depended on the accuracy of the velocity measurement, the correlation coefficients and SNRs of less than 90% and 30 dB, respectively, were eliminated, and suspected spikes were removed using the phase-space threshold method [18]. The distance between the velocity of each layer and the bed surface was automatically measured by the instrument, which greatly reduced the error compared to that which would have resulted from manual measurement. This provided a more reliable and effective method for analyzing the velocity distribution near the bed than single-point ADV.

3. Methods

In the flow field of the bend considered in this experiment, there was transverse velocity. The bed shear stress obtained using the Reynolds method consisted of two parts: streamwise shear stress ($-\rho(\overline{u'w'})$) and transverse shear stress ($-\rho(\overline{v'w'})$). Similarly, the bed shear stress obtained by the log-law method consisted of the streamwise shear stress calculated from the time-averaged velocity in the streamwise direction (s) and the transverse shear stress calculated from the time-averaged velocity in the transverse direction (n). Because the flow field in this experiment belonged to a strong three-dimensional complex flow field, it was difficult to obtain the extrapolated values for the bed by the Reynolds, TKE, and TKE-w′ methods. Therefore, the bed shear stress was calculated by measuring the corresponding maximum value near the bed.

Figure 3a,b shows the shear stress distributions along the test line (n/L = 1.4, S29) according to the TKE method (Equation (1)) and TKE-w′ method (Equation (2)), respectively. The bed shear stress was the peak-value at about z_a = 0.05 m above the bed.

Figure 3c,d gives the Reynolds shear stress distributions in the streamwise (s) and transverse (n) directions, respectively, calculated by the Reynolds method (Equation (3)) along the test line (n/L = 1.4, $S29$).

The two directions of shear stress increased rapidly from the bottom of the bed and reached the peak-values at z_a = 0.015 m and z_a = 0.010 m, respectively. As shown in Figure 3a, with the increasing distance from the bed ($0 < z_a < 0.05$), the turbulence of the flow increased rapidly. When the distance from the bed was bigger than 0.05 m, the turbulence of the flow gradually decreased. Figure 3e,f shows the velocity distribution along test lines (n/L = 1.4, $S29$ and n/L = 4.8, $S42$, respectively). In general, streamwise velocity (u) was more in accordance with a logarithmic distribution than the transverse velocity (v). The transverse velocity (v) was only consistent with a logarithmic distribution in the vicinity of the bed. The values obtained for the coefficient of determination R^2 of the logarithmic fitting formula for various cases were greater than 0.8. Using the fitting formula and Equation (4), the bed shear stress at the corresponding measuring point can be calculated.

The positive Reynolds stress ($-\rho(\overline{u'w'})$) of the liquid particle corresponded to the positive time-averaged velocity gradient ($\frac{d\overline{u}}{dz}$), and the negative Reynolds stress corresponded to the negative time-averaged velocity gradient. According to Prandtl's mixing length theory, the Reynolds stress value was related to the time-averaged velocity gradient as follows:

$$-\rho(\overline{u'w'}) = \rho l^2 \left(\frac{d\overline{u}}{dz}\right)^2 \quad (6)$$

where l is the mixing length, $l = \kappa y$; κ is a dimensionless constant. Figure 3d,e shows that during $0 < z_a < 0.025$, the positive velocity gradient corresponded to the positive Reynolds stress, and during $0.025 < z_a < 0.120$, the negative velocity gradient corresponded to the negative Reynolds stress. During $0.12 < z_a < 0.17$, the velocity gradient was small, and its corresponding Reynolds stress value was basically zero.

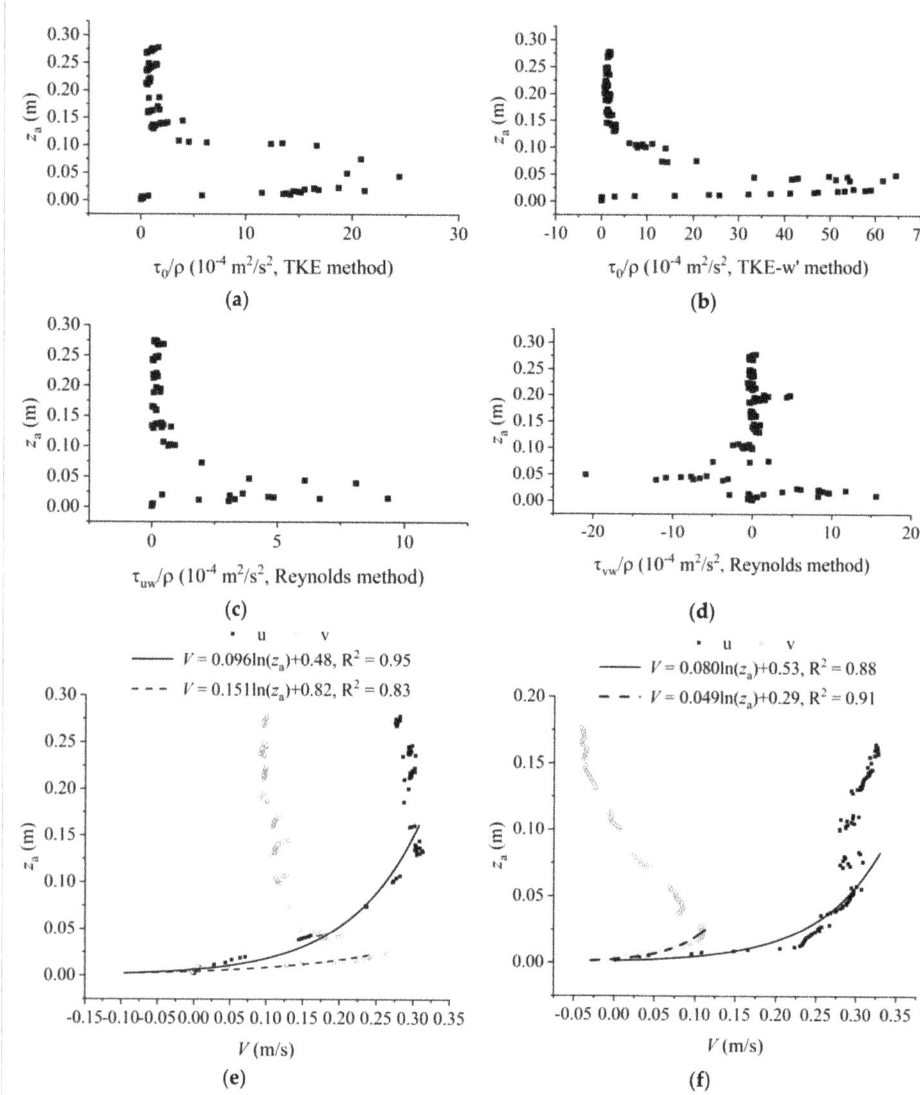

Figure 3. Bed shear stress estimates: (**a**) TKE method ($n/L = 1.4$, *S29*); (**b**) TKE-w′ method ($n/L = 1.4$, *S29*); (**c**) Reynolds method ($n/L = 1.4$, *S29*); (**d**) Reynolds method ($n/L = 1.4$, *S29*); (**e**) log-law method ($n/L = 1.4$, *S29*); (**f**) log-law method ($n/L = 4.8$, *S31*).

4. Results and Discussion

4.1. Comparison of Methods for Bed Shear Stress Estimates in Selected Sections

Figure 4 shows a comparison of the four methods for bed shear stress estimates in selected sections. The values estimated according to the four methods were basically the same in trend and value. The entrance section *S0* of the bend was not affected by the groyne and could be used for analysis of the bed shear stress in a simple flow field. The trends of the four types of bed shear stress estimates along section *S0* were basically the same, reaching a maximum at $n/L = 2.4$. The measurement values according to each method changed slightly along the section, as section *S0* was at the entrance of the bend, and the influence of the bend on the flow field was still weak at that point. The bed shear stress in the middle of the flume was slightly greater than that on the banks because of the retarding effect of the side wall of the flume on the flow. In section *S*0, the results obtained with the TKE method and Reynolds method were the most similar, followed by those obtained with the log-law method, while the values obtained with the TKE-w′ method were the largest of those obtained with the four methods.

In regions *S*29–*S*33, there was a strong three-dimensional flow field because of the combined effect of the bend and the groyne. On the upstream side of groyne (*S*29–*S*29.5), the bed shear stress on the outer side ($1 < n/L < 3$) was significantly larger than that on the inner side ($3 < n/L < 6$). The values for the flow fields in each section, obtained using the TKE-w′ method, were typically larger than those obtained with the log-law, TKE, and Reynolds methods, but there was little difference between the values obtained with the other three methods for the inner side ($3 < n/L < 6$), where the bed shear stress level was relatively low. The distribution of the bed shear stress in section *S*29 was similar to that in section *S*29.5. The bed shear stress in section *S*29.5 was lower than that in section *S*29 in the range of $0 < n/L < 1$ because the region ($0 < n/L < 1$, *S*29.5) was within the wing of the groyne, and the flow turbulence and velocity were suppressed. Sections *S*30.5–*S*32 were on the downstream side of the groyne. The bed shear stress in the middle of the flume ($1.2 < n/L < 4$) decreased gradually in the flow direction, and the bed shear stress in the region of the groyne wing ($0 < n/L < 1$) increased gradually. This is because, with increasing distance from the groyne, the influence of the groyne on the flow turbulence and velocity in the middle of the flume decreased, and the bed shear stress was lower. However, the restraint effect of the groyne on the flow turbulence and velocity inside the groyne wing gradually decreased, and the bed shear stress increased accordingly. In section S33, aside from the values obtained with the TKE-w′ method in the range of $0 < n/L < 1$, the bed shear stress for whole sections tended to be stable, and the bed shear stress on the inner side of the groyne wing ($0 < n/L < 1$) decreased, indicating that the distribution of the downstream bed shear stress became progressively less influenced by the groyne.

Despite errors in the measurement of the Reynolds stress, the Reynolds method is considered to be relatively reliable [6,14,19]. In Figure 5, the TKE method, TKE-w′, and log-law methods were fitted linearly with a zero intercept, with slopes of 0.836 ($R^2 = 0.070$), 2.035 ($R^2 = 0.673$), and 0.841 ($R^2 = 0.290$), respectively. The slope of the solid line is 1. The slope of the log-law method was closest to 1, as shown in Figure 5, indicating that it was closer to the Reynolds method values at both low and high stress levels (Figure 5). The slope of the TKE-w′ method was much greater than 1, as shown in Figure 5, which indicated that the stress values were significantly greater than the Reynolds method values for both low and high stress levels. The slope of the TKE method was close to that of the log-law method, but its R^2 was the smallest, indicating the greatest dispersion, which was higher than that of the Reynolds method at low stress levels and lower than that of the Reynolds method at high stress levels (Figure 5).

Figure 6a shows topographic illustrations of bed scouring around the groyne (positive values represent silting; negative values represent scouring). The scouring depth was the highest at the upstream head of the groyne, while sediment deposits occurred at the inner bank. Figure 6b shows the velocity distribution near the groyne. The velocity vectors indicate the depth-averaged velocity vectors (UV components) and depth-averaged vertical velocity-filled contours (W component, positive values represent the upward velocity). Figure 6c–f shows the bed shear stress contours (using the Kriging method) for the TKE, TKE-w′, Reynolds, and log-law methods, respectively. There were two high bed shear stress zones identified by the four methods. The first one was located on the upstream side near the groyne, where both the horizontal and vertical velocities were large; the second one was located on the downstream side at a distance from the groyne, where the horizontal velocity was lower than in the surrounding area, but the vertical velocity was higher than in the surrounding area, so the vertical velocity in this area contributed greatly to the formation of high bed shear stress. The bed shear stress in the second area was smaller than that in the first area, mainly because of the lower horizontal velocity (Figure 6b). Chrisohoides et al. (2003) and Biron et al. (2004) assumed that there was a strong correlation between the scouring area and high bed shear stress [14,20]. The four types of bed shear stress distribution contours identified in this study had certain similarities with the topographic contours. High bed shear stress corresponded to scouring area, but the bed shear stress in the deepest scouring area enclosed by the side wall and groyne wing on the upstream side of the groyne was relatively small and was mainly located in the back-flow zone with low velocity (Figure 6b).

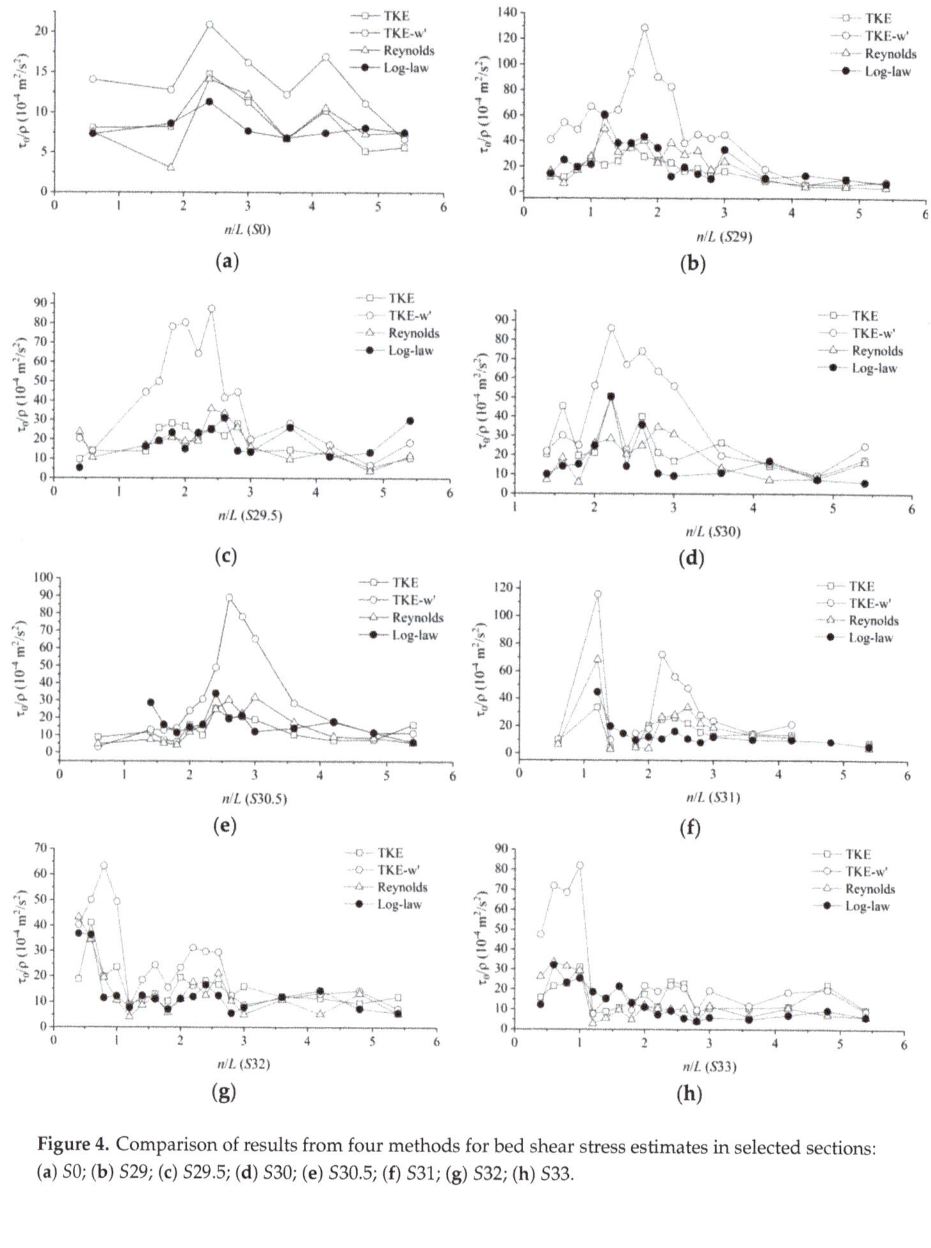

Figure 4. Comparison of results from four methods for bed shear stress estimates in selected sections: (**a**) *S*0; (**b**) *S*29; (**c**) *S*29.5; (**d**) *S*30; (**e**) *S*30.5; (**f**) *S*31; (**g**) *S*32; (**h**) *S*33.

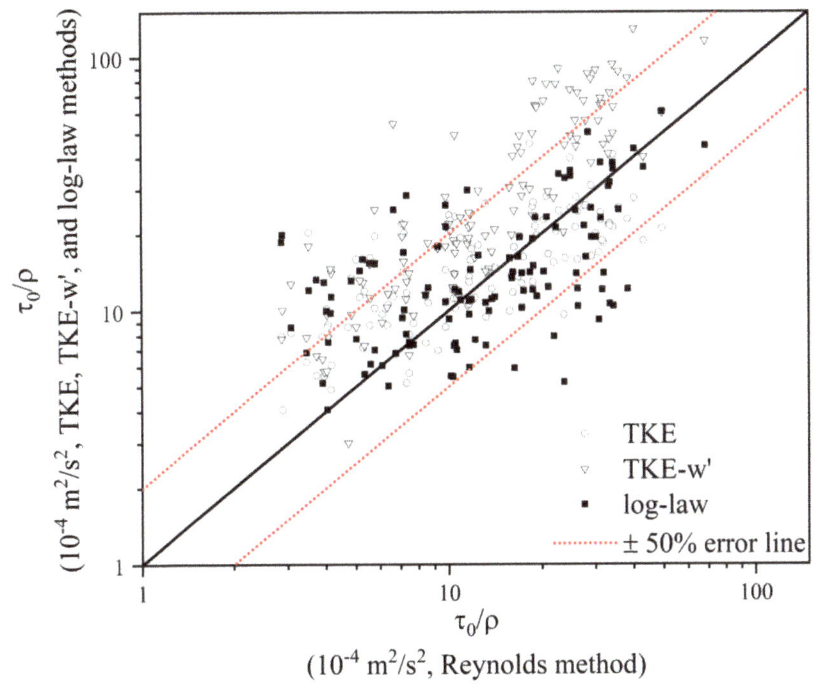

Figure 5. Comparison of Reynolds method with the TKE, TKE-w′, and log-law methods.

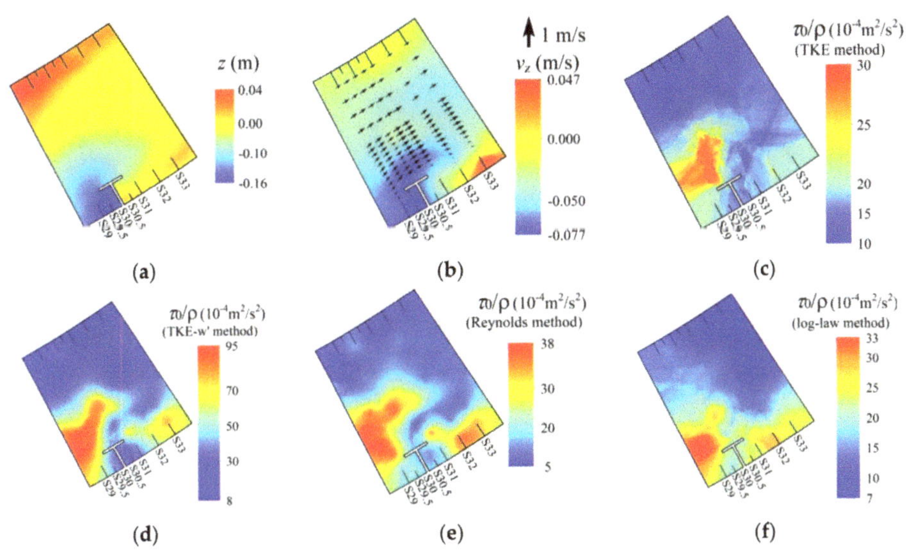

Figure 6. (a) Bed topography around the groyne; (b) depth-averaged velocity vectors (UV components, "1 m/s" corresponds to the size of the arrow on the left) and depth-averaged vertical velocity filled contours (W component, positive values represent upward velocity); (c) bed shear stress (TKE method); (d) bed shear stress (TKE-w′ method); (e) bed shear stress (Reynolds method); (f) bed shear stress (log-law method).

4.2. Re-Estimation of Parameters of TKE and TKE-w′ Methods

The values of the constants (C_1 and C_2) of the TKE and TKE-w′ methods may vary dramatically near the bed, so they can be re-estimated based on measurements obtained in the laboratory or in natural rivers [14].

According to the calculation results presented in the previous section (Figure 5), the values of the constants C_1 and C_2 were adjusted to 0.23 and 0.44, and the bed shear stress values according to the TKE and TKE-w′ methods were recalculated. Figure 7 shows linear regressions with zero intercepts using the TKE and TKE-w′ methods, with slopes of 1 (R^2 = 0.070) and 1 (R^2 = 0.673), respectively. The slope of the solid line is 1. After adjusting the parameter values, the slopes of the regressions for the TKE method and the TKE-w′ method were 1, but the R^2 for the former method was much smaller than that for the latter, indicating that there was much more scatter in the TKE results, probably due to the higher error associated with the horizontal components of velocity [14]. Based on the experimental results obtained in this study, the TKE-w′ method was judged to be more suitable for use in bed shear stress estimations than the TKE method, but the value of its constant C_2 needed to be re-estimated. In this study, the re-estimated values of constants C_1 and C_2 varied by 21.1% and 51.1%, respectively, compared with the original values. Compared with the TKE-w′ method, the variation in the value of the TKE method's constant C_1 was smaller, and the slope of the regression for the Reynolds method was closer to 1 (Figure 5) with no change in the constant. Therefore, when the constant values were not re-estimated or could not be estimated, the TKE method may yield more reasonable bed shear stress estimates than the TKE-w′ method. More experiments are needed to verify whether the TKE method performs better under other experimental conditions.

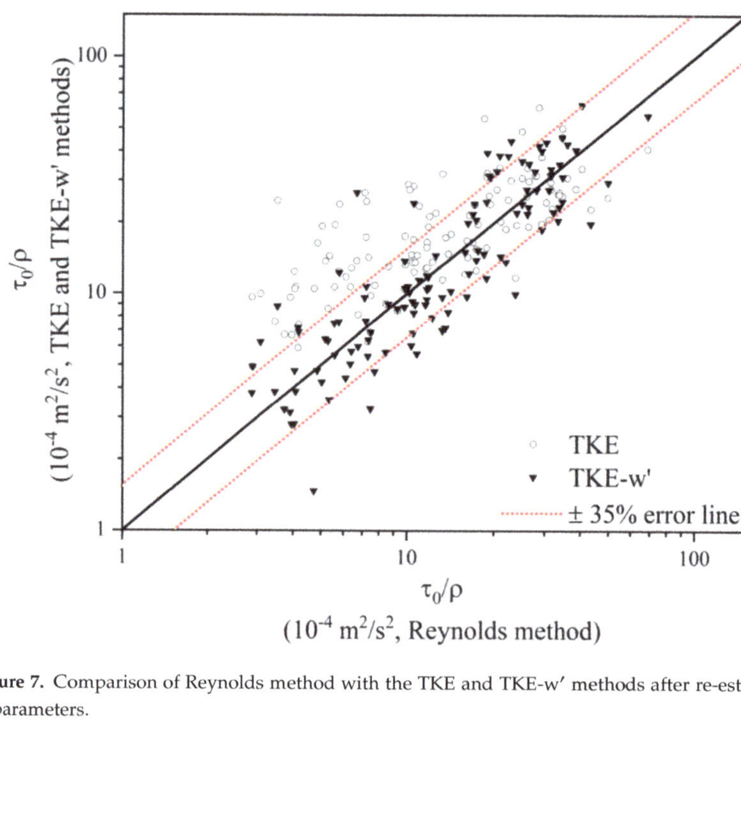

Figure 7. Comparison of Reynolds method with the TKE and TKE-w′ methods after re-estimation of parameters.

4.3. Transverse Bed Shear Stress

The overall magnitude of the bed shear stress is often of greater concern than the magnitudes of the directional components of the bed shear stress, especially the transverse bed shear stress. However, the transverse transport of sediment is closely related to the value of the transverse bed shear stress [21]. Figure 8a,b shows the transverse bed shear stress estimated by the Reynolds method and the log-law method, respectively. In Figure 8a,b, the difference is small in the high shear stress region, but large in the low shear stress region. Therefore, it is difficult to apply the logarithmic law to determine the transverse shear stress under the condition of low shear stress level (usually corresponding to low transverse velocity). The maximum transverse bed shear stress occurred near the groyne wing, upstream of the groyne, because the flow was blocked by the groyne and the streamlines were deflected, resulting in high transverse velocity (Figure 6b). Because of the high transverse bed shear stress, transverse sediment transport occurred, leading to the formation of large transverse scouring zones (Figure 6a).

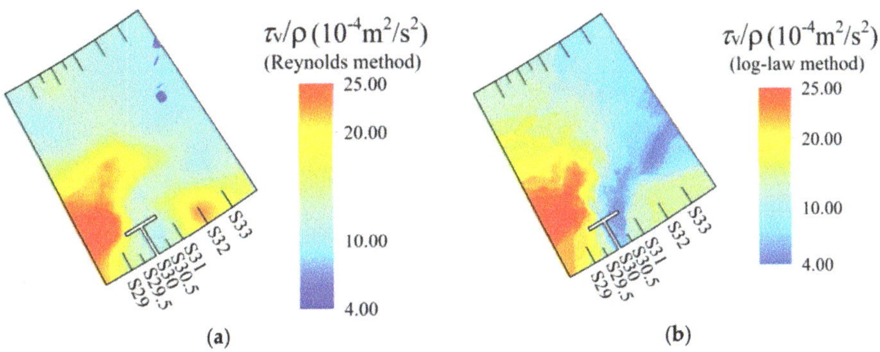

Figure 8. Transverse bed shear stress estimates: (**a**) Reynolds method; (**b**) log-law method.

Both the TKE and TKE-w′ methods yielded direct estimates of the bed shear stress, whereas the Reynolds method and the log-law method could be used less directly to estimate the streamwise bed shear stress and transverse bed shear stress, which correspond to the streamwise and transverse transport of the sediment, respectively, and are of great significance in the study of fluvial processes.

4.4. Height Analysis for Single-Point Measurements

In experiments or in observations of natural rivers, it is often very difficult to measure the velocity of full-flow-depth profiles. Single-point measurements near the bed are often used to estimate the bed shear stress. Therefore, determination of the height at which single-point measurements should be taken is an important part of estimating the bed shear stress. In fluvial studies, it is generally considered that a point close to the bed but above the roughness layer is ideal for estimating the average bed shear stress on a roughness scale (e.g., Babaeyan-Koopaei et al. [5]). Simple boundary layer flow concepts confirm that both the Reynolds shear stress and turbulent energy vary with the height above the bed, reaching a maximum value when the height of the bed is 0.1 times the depth of the water [22]. Some researchers have obtained peak-values at the same dimensionless distance (0.1) above gravel beds [6]. The applicability of this dimensionless distance 0.1 (e.g., for different bed surface roughnesses, and taking into consideration parameters of the measurement equipment, the influence of velocity, etc.) and the optimal measurement height, in terms of the absolute height above the bed or the relative height, requires further research [14]. Figure 9 shows a comparison of the peak-value heights corresponding to the TKE method, the TKE-w′ method, and two directions (s and n) of the Reynolds method in section S0. Because the S0 section was at the entrance of the bend, where the topography changed slightly, the flow depth H of each point in the section remained basically unchanged at a value of $H = 0.24$ m. The peak-value heights for the TKE, TKE-w′, and Reynolds methods were not significantly different at all points in the section and varied slightly across the section. This can be expressed by a uniform height $z_a = 0.013$ m ($z_a/H = 0.05$).

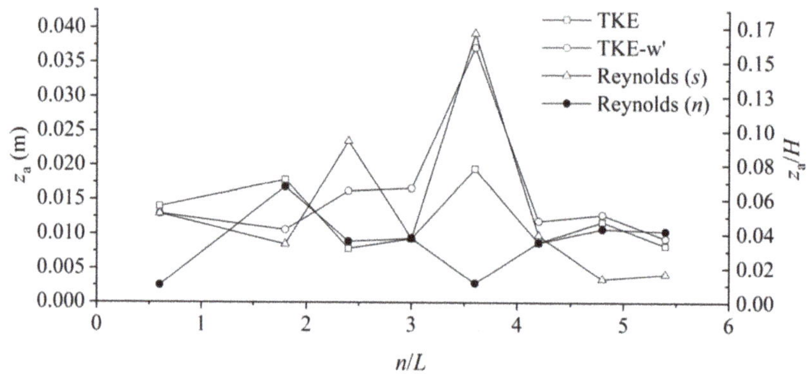

Figure 9. The peak-value heights of section *S*0.

Figure 10 shows a comparison of the peak-value heights for the TKE method, TKE-w′ method, and Reynolds method (*s* and *n*) near the groyne. In this zone, because of the considerable change in the strong three-dimensional flow structure and topography, the peak-value height changes for the TKE, TKE-w′, and Reynolds methods were more complicated than those for section *S*0, but they still had certain similarities in terms of their spatial variation. The absolute height distribution ranges (the difference between the maximum and the minimum) according to the TKE method, TKE-w′ method, and Reynolds method (streamwise and transverse) were 0.032, 0.034, 0.023, and 0.023 m, respectively, and the relative height distribution ranges were 0.12, 0.16, 0.11, and 0.15, respectively. The absolute height distribution range and the relative height distribution range of the TKE-w′ method were the largest among the three methods, and the peak-value height was the most unstable of the three methods. The absolute height distribution range (0.023 m) and the relative height distribution range (0.11) of the streamwise Reynolds method were the smallest of the three methods, but the range was still sufficiently large. Therefore, in a complex, strong, three-dimensional flow field, the height of the measurement point (relative or absolute) should be re-estimated when the single-point measurement is used to estimate the bed shear stress.

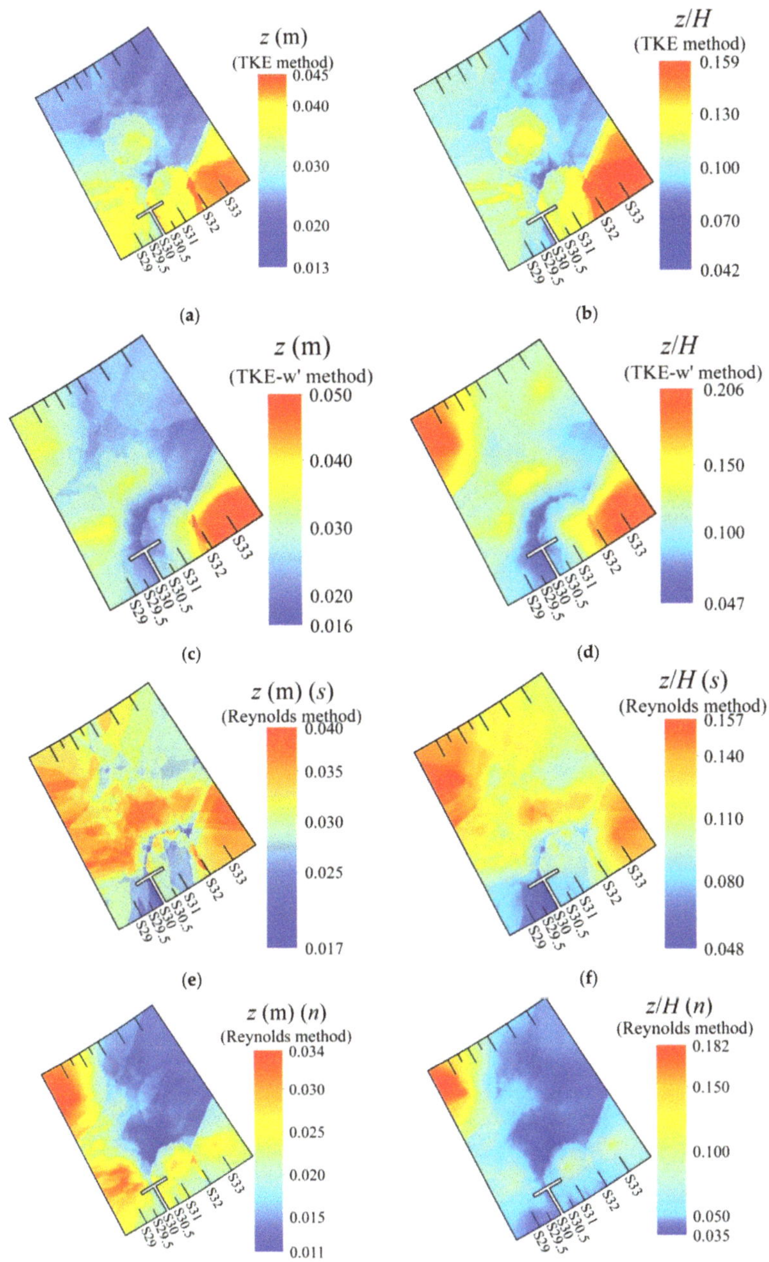

Figure 10. Peak-value heights near the groyne: (**a**,**b**) TKE method; (**c**,**d**) TKE-w′ method; (**e**,**f**) Reynolds method (s); (**g**,**h**) Reynolds method (n).

5. Conclusions

In this study, the complex flow field in a 60° bend of a river with a groyne was examined, and the bed shear stress was estimated by the turbulent energy method (TKE method), modified turbulent energy method (TKE-w′ method), Reynolds stress method (Reynolds method), and logarithmic law (log-law method). The results are summarized as follows.

(1) It is feasible to estimate the bed shear stress in a complex flow field using velocity data for the log-law region near the bed.
(2) In this study, river bed shear stress estimates obtained using the log-law method were found to be closer to those obtained using the Reynolds method at both low and high shear stress levels than those obtained using the TKE and TKE-w′ methods.
(3) If the turbulent kinetic energy method was used to calculate the bed shear stress, the experimental results showed that the TKE-w′ method was more suitable for bed shear stress estimation than the TKE method, but the value of its constant C_2 needed to be re-estimated.
(4) The Reynolds method and the log-law method can both be used to determine the streamwise and transverse directions of bed shear stress.
(5) In a complex, strong, three-dimensional flow field, the height of the measurement point (relative or absolute) should be re-estimated when the single point measurement is used to estimate the bed shear stress.

Author Contributions: Conceptualization, L.Z. and F.Z.; methodology, L.Z.; software, L.Z.; validation, L.Z. and F.Z.; formal analysis, L.Z.; investigation, L.Z., F.Z., A.C., Z.S. and S.T.; resources, L.Z. and F.Z.; data curation, L.Z. and F.Z.; writing—original draft preparation, L.Z.; writing—review and editing, L.Z. and F.Z.; visualization, L.Z.; supervision, F.Z.; project administration, F.Z. All authors have read and agreed to the published version of the manuscript.

Funding: This research was funded by the National Natural Science Foundation of China, grant number 51679157, the National Key Research and Development Program, grant number 2016YFC0401705, and the Sichuan Science and Technology Program, grant number 2019JDTD0007.

Acknowledgments: The authors would like to thank Wangru Wei for his helpful advice and discussion about this paper.

Conflicts of Interest: The authors declare no conflict of interest.

Abbreviations

H	Approaching flow depth
L	Groyne length (0.1 m)
Q	Total discharge
R_c	Radius of the centerline of the bend
U_0	Velocity of the approaching flow
W	Cross-sectional width of the main channel
d_{50}	Median size of the sediment
h	Wing height of the groyne above the bed
i	Flume longitudinal slope (0.001)
l	The mixing length
n	Transverse location coordinates
s	Longitudinal location coordinates
u, v, w	Velocity of the streamwise, transverse, and vertical components
u', v' and w'	Velocity fluctuations of the streamwise, transverse, and vertical components
$\overline{u}$, $\overline{v}$ and $\overline{w}$	Time-averaged velocity of the streamwise, transverse, and vertical components
u_*	Friction velocity
x, y, z	Rectangular Cartesian coordinates
z_a	Height above the bed
z_0	Statistical roughness length
v_c	Critical velocity of the sediment

φ	Central angle of the main channel curved segments
κ	Von Karman's constant (0.4)
τ_0	Bed shear stress
τ_u and τ_v	Bed shear stress of the streamwise and transverse components

References

1. Wilcock, P.R. Estimating local bed shear stress from velocity observations. *Water Resour. Res.* **1996**, *32*, 3361–3366. [CrossRef]
2. Pujara, N.; Liu, P.L.F. Direct measurements of local bed shear stress in the presence of pressure gradients. *Exp. Fluids* **2014**, *55*. [CrossRef]
3. Kim, S.C.; Friedrichs, C.T.; Maa, J.P.Y.; Wright, L.D. Estimating bottom stress in tidal boundary layer from Acoustic Doppler Velocimeter data. *J. Hydraul. Eng. ASCE* **2000**, *126*, 399–406. [CrossRef]
4. Voulgaris, G.; Trowbridge, J.H. Evaluation of the Acoustic Doppler Velocimeter (ADV) for turbulence measurements. *J. Atmos. Ocean. Technol.* **1998**, *15*, 272–289. [CrossRef]
5. Babaeyan-Koopaei, K.; Ervine, D.A.; Carling, P.A.; Cao, Z. Velocity and turbulence measurements for two overbank flow events in River Severn. *J. Hydraul. Eng. ASCE* **2002**, *128*, 891–900. [CrossRef]
6. Nikora, V.; Goring, D. Flow turbulence over fixed and weakly mobile gravel beds. *J. Hydraul. Eng. ASCE* **2000**, *126*, 679–690. [CrossRef]
7. Chen, X.W.; Chiew, Y.M. Response of velocity and turbulence to sudden change of bed roughness in open-channel flow. *J. Hydraul. Eng. ASCE* **2003**, *129*, 35–43. [CrossRef]
8. Heathershaw, A.D. Turbulent structure of the bottom boundary-layer in a tidal current. *Geophys. J. R. Astron. Soc.* **1979**, *58*, 395–430. [CrossRef]
9. Bin Asad, S.M.S.; Lundström, T.S.; Andersson, A.G.; Hellström, J.G.I.; Leonardsson, K. Wall shear stress measurement on curve objects with PIV in connection to benthic fauna in regulated rivers. *Water* **2019**, *11*, 650. [CrossRef]
10. Hrissanthou, V.; Hartmann, S. Measurements of critical shear stress in sewers. *Water Res.* **1998**, *32*, 2035–2040. [CrossRef]
11. Williams, J.J. Drag and sediment dispersion over sand waves. *Estuar. Coast. Shelf Sci.* **1995**, *41*, 659–687. [CrossRef]
12. Liu, C.J.; Li, D.X.; Wang, X.K. Friction velocity and velocity distribution for uniform flow in open channel. *J. Hydraul. Eng.* **2005**, *130*, 950–955. (In Chinese) [CrossRef]
13. Shivpure, V.; Sharm, A.; Kumar, B. Comparison of bed shear stress in plane and curvilinear bed channel using multiple criteria. *Water Resour.* **2016**, *43*, 79–85. [CrossRef]
14. Biron, P.M.; Robson, C.; Lapointe, M.F.; Gaskin, S.J. Comparing different methods of bed shear stress estimates in simple and complex flow fields. *Earth Surf. Process. Landf.* **2004**, *29*, 1403–1415. [CrossRef]
15. Izquierdo, U.; Esteban, G.A.; Blanco, J.M.; Albaina, I.; Peña, A. Experimental validation of a CFD model using a narrow wave flume. *Appl. Ocean Res.* **2019**, *86*, 1–12. [CrossRef]
16. Zhang, R.J. *River Dynamics*; Wuhan University Press: Wuhan, China, 2007.
17. Zhang, L.Y.; Zhang, F.X.; Cai, A.L.; Song, Z.M.; Tong, S.L. Application of 3D laser scanning technology in simulating river flows. *Water Resour. Power* **2020**, *38*, 121–124. (In Chinese)
18. Goring, D.G.; Nikora, V.I. Despiking Acoustic Doppler Velocimeter data. *J. Hydraul. Eng. ASCE* **2002**, *128*, 117–126. [CrossRef]
19. Poggi, D.; Porporato, A.; Ridolfi, L. Analysis of the small-scale structure of turbulence on smooth and rough walls. *Phys. Fluids* **2003**, *15*, 35–46. [CrossRef]
20. Chrisohoides, A.; Sotiropoulos, F.; Sturm, T.W. Coherent structures in flat-bed abutment flow: Computational fluid dynamics simulations and experiments. *J. Hydraul. Eng.* **2003**, *129*, 177–186. [CrossRef]

21. Kiraga, M.; Popek, Z. Bed shear stress influence on local scour geometry properties in various flume development conditions. *Water* **2019**, *11*, 2346. [CrossRef]
22. Song, T.; Chiew, Y.M. Turbulence measurement in nonuniform open-channel flow using Acoustic Doppler Velocimeter (ADV). *J. Eng. Mech. ASCE* **2001**, *127*, 219–232. [CrossRef]

water

MDPI

Article

Optimal Strategy to Tackle a 2D Numerical Analysis of Non-Uniform Flow over Artificial Dune Regions: A Comparison with Bibliography Experimental Results

Jungkyu Ahn [1], Jaelyong Lee [1] and Sung Won Park [2,*]

1 Department of Civil and Environmental Engineering, Incheon National University, Incheon 22012, Korea; ahnjk@inu.ac.kr (J.A.); dlwofydsla1@naver.com (J.L.)

2 Department of Data-centric Problem-Solving Research, Korea Institute of Science and Technology Information, Daejeon 34141, Korea

* Correspondence: swpark@kisti.re.kr; Tel.: +82-42-869-1624

Received: 8 July 2020; Accepted: 14 August 2020; Published: 19 August 2020

Abstract: Flow simulation over a dune requires the proper input of roughness coefficients. This study analyzed a numerical simulation of open-channel turbulent flow over two-dimensional fixed dunes to reveal the effect of roughness on the dune bottom, and to determine the optimized combination of the turbulence scheme and the roughness height formula. The most appropriate roughness values and turbulence models were applied using Reynolds-averaged Navier–Stokes models. Seven methods were chosen to estimate the bed roughness properties at the inlet boundary section. The results of all cases calculated with the OpenFOAM toolbox were compared with laboratory experimental data for model validation. The performances of all bed roughness variations were evaluated according to the stream-wise and depth-wise directions with nondimensional values. Consequently, it was revealed that the combination of bottom roughness length scale at the inlet boundary and the k-ω shear-stress transport (SST) model was the most suitable for the flow separation zone and turbulent properties near the channel bottom.

Keywords: bed roughness; open-channel flow; OpenFOAM; Reynolds-averaged Navier–Stokes model; turbulence model; two-dimensional dune

1. Introduction

Bed forms such as ripples and dunes in an alluvial river change naturally with the interaction between the river and sediment. In particular, because of scour around hydraulic structures such as piers and dams, bottom protection of weirs is considered to be an important phenomenon of sediment transport, since the excessive loss of bed materials can have severe effects and cause damage to structures [1–4]. In general, turbulent flow with arbitrary flow characteristics is considered to be the predominant cause of scouring. This process is too complicated to predict because it is a result of the relationships among the water flow around structures and the irregular bed properties in natural rivers. Analysis of the characteristics of turbulent flow in rivers requires interaction between the flow and the channel bed, and one of the most important processes for the numerical analysis is the determination of the bed roughness.

Most numerical models of local scour, sediment transport, and deposition separately use an uncoupled scheme, turbulence, and sediment transport [5]. The flow computation is conducted before the sediment computation. Then, the bed shape is updated and processed to the next time step. In the present study, we analyzed turbulent structures in flow over fixed dunes to find the best method for turbulent computation. Among the various irregular bed shapes in alluvial rivers, the dune is one of the most interesting in context of understanding its interaction. Numerous experimental approaches on this topic have been carried out [6–16]. Particularly, it has been revealed that dunes were formed by multiple mechanisms resulting from the complex combination of bedform, grain motions, and flow dynamics that evolve as a bed develops [17]. Five major regional flow characteristics are summarized in Figure 1 based on previous researches [15,17]. Therefore, this study focused on the effects of bed roughness on turbulent flow, which is one of the factors influencing flow in natural streams.

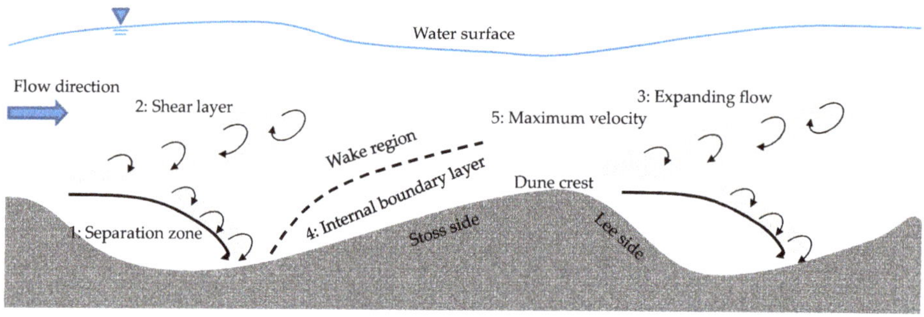

Figure 1. Schematic diagram of principal regions of flow over the dunes [15,17].

Physical and numerical approaches can be applied to investigate bed roughness and its effects. The physical approach requires no parameters to start modeling. However, it is very difficult to maintain constant and stable conditions during modeling, especially for large scale models. In addition, measuring error is inevitable. A considerable disadvantage is the scale effect. Therefore, numerical simulation is widely applied because it minimizes the scaling effects and facilities [18–25]. Although adequate results can be obtained with this approach, there are many limitations owing to the combination of complex properties of the turbulent flow. Therefore, the appropriate combination of the model parameters and the turbulent modeling technique is essential.

Most numerical simulations of turbulent flow with various turbulence schemes have been studied to solve the Navier–Stokes (NS) equation to capture the effect of turbulence motion. Direct numerical simulation (DNS) is a simulation in computational fluid dynamics (CFD) for solving NS equations of turbulent flow without additional turbulence schemes. This method is the most accurate for modeling flow characteristics. However, the computational cost of DNS is high because all spatial scales of the turbulence must be resolved in a mesh from the minimum dissipated scale, and it is difficult to simulate for large, complex applications [26–28]; nonetheless, turbulence modeling is conducted in practical engineering applications. To overcome the diseconomy of the DNS method in terms of computational time and effort, large eddy simulation (LES) methods with lower computational costs and the Reynolds-averaged Navier–Stokes (RANS) methods based on averaging the flow equations yielding the Reynolds-Averaged Navier–Stokes (RANS) equations were applied with accurate flow simulations [28,29]. The closure problem was solved with transport equations to reproduce the behavior of turbulent flow, and then the turbulence scales were related to a turbulent viscosity. Two-equation models provided a full description of turbulence in terms of length and time scales; thus, they facilitated the reproduction of various flow patterns [26,28].

In this study, we used the turbulence model based on the RANS equations to simulate the numerical modeling for turbulence flow over a fixed dune, and to reduce the computational time by solving the mean velocity and effects of the temporal fluctuation. Numerous numerical studies have provided reasonable results for turbulent flow on sandy dunes using various RANS models [2,23,30–36]. Considering the previous studies, we selected k-ε and k-ω shear-stress transport (SST) turbulence models from the RANS models. In addition, a two-dimensional (stream-wise and depth-wise direction) model was applied for flow and turbulence analysis over fixed dunes because these have been considered to be the dominant factors of the maximum scour depth in the design criteria of hydraulic structures in previous researches [1,4,37–40].

In addition, bed roughness, which depends on the size of the bed materials, is one of the most important factors in open-channel flow. This property is presented through various suggestions based on previously suggested experimental data and is expressed as the grain size of the bed material. In the general type of numerical modeling, properties of bed materials are chosen according to empirical formula. Representatively, several identical dunes were studied to identify the effect of the dune wall roughness [35] and dimensions by using the standard k-ε model and the Spalart–Allmaras model [41] with one-equation models and by applying the Colebrook–White formula [42] to set the roughness value of the wall. In addition, previous studies [8,43] have shown the results of various hydrodynamic factors, i.e., the time-averaged velocity, eddy viscosity, free-surface elevation, wall stress, wall pressure, friction, and form resistance data, to determine a suitable formula for bed roughness. A numerical analysis was conducted to investigate the form drag of several idealized dune configurations and to provide various flood wave models by using the Nikuradse roughness parameter from previous research [44,45]. Their predictions of the velocity profiles and flow separation zones were in general agreement with the experimental data from previous research [46]; however, only the standard k-ε model was used. Therefore, the objective of this study was to apply several combinations of empirical formula for the bed roughness in the numerical approaches for the fixed dune, and to determine the most suitable formula. Six experimental formula were used for turbulent distribution on the bed of the artificial dune using the data from previous research [6].

The physical experimental data of dune profiles and conditions were obtained from the previous research [6]; these data have been widely used and proven to be accurate in previous studies [2,23,32,35,36,47,48]. We employed the OpenFOAM (The OpenFOAM Foundation Ltd. London, United Kingdom) toolbox, an open-source CFD platform based on a cell-centered finite volume method. The simulated results of flow velocity and turbulent kinetic energy were obtained, and the optimal best combinations were determined by comparing the results with those of physical modeling [6]. The goodness-of-fit was quantified by root mean square error (RMSE).

2. Numerical Method

The OpenFOAM toolbox is a free CFD program developed by OpenCFD based on C++, a collection of solvers capable of simulating the flow of various fluids, and open code software. This toolbox is basically a three-dimensional model that solves the partial differential equation based on the finite-volume method and numerically presents the governing equation of fluid motion. The governing equation for incompressible flow is the RANS equation using the Reynolds-averaged theorem. The Reynolds stresses are modeled based on the eddy viscosity. The continuity equation and the momentum equation in vector form are as follows:

$$\nabla \cdot \vec{u} = 0 \tag{1}$$

$$\frac{\partial \vec{u}}{\partial t} + (\vec{u} \cdot \nabla)\vec{u} = -p + \nabla \cdot \left[(\nu + \nu_T)(\nabla \vec{u} + \nabla \vec{u}^T) \right] \tag{2}$$

where $\vec{u}$ is the velocity vector, and upper T denotes the transposed gradient, p is the pressure, ν is the kinematic viscosity coefficient, and ν_T is the turbulent kinematic eddy viscosity coefficient which is supposed to simulate the effect of unsolved velocity fluctuations.

In the OpenFOAM toolbox, the standard k-ε turbulence model [49] is one of the most commonly used turbulence models to calculate flow properties for turbulent flow. It describes turbulence by means of two partial differential equations (PDEs) for the following two variables: turbulent kinetic energy (TKE), k ($= 0.5\sqrt{\overline{u'^2} + \overline{v'^2} + \overline{w'^2}}$, where u', v', and w' denote the three-dimensional velocity fluctuation from the time-averaged velocity data); and the turbulent dissipation, ε. This model is robust, economical, and relatively accurate for many cases [40]. It is designed as a high Reynolds number model and provides a relatively reasonable formula, especially in fully developed turbulent flow; it is simple to implement and performs relatively well for boundary flow. However, the standard model fails to properly predict flows close to the wall region where the Reynolds number is low, because the eddy viscosity is overestimated by the two equations in this case [44,50]. Therefore, the wall function was applied to improve the calculation accuracy near the wall; it was applied to the first cell of the wall boundary, thus applying a constant roughness to these walls. The turbulent viscosity can be determined by two convention-diffusion-reaction equations for computation, presented as follows:

$$\frac{\partial k}{\partial t} + \nabla \cdot (\vec{u}k) - \nabla \cdot (\nu + \nu_T)\nabla k = P_k - \frac{\varepsilon}{k} \tag{3}$$

$$\frac{\partial \varepsilon}{\partial t} + \nabla \cdot (\vec{u}\varepsilon) - \nabla \cdot (\nu + \nu_T)\nabla \varepsilon = (C_1 P_k - C_2)\frac{\varepsilon}{k} \tag{4}$$

where P_k is the turbulent viscosity production due to viscosity. In addition, ν_T is defined as:

$$\nu_T = \frac{C_\mu \sqrt{k}}{\varepsilon} \tag{5}$$

where C_1, C_2, and C_μ are the empirical constants with values of 1.44, 1.92, and 0.09, respectively.

The k-ω SST model is a two-equation model developed by [50] and is similar to the k-ω model [51]. The k-ω SST model effectively combines the formulations of the k-ε and k-ω models. The model assumes that the turbulent viscosity is related to values of k and ω. The k-ε model is similar to the k-ω model but it is more accurate for wall effect treatment. However, the SST model avoids the problem of the k-ω model's sensitivity to inlet properties in freestream, by switching the formula of both models to each other in freestream [50]. Therefore, the turbulence characteristics near the wall and the freestream region can be predicted appropriately, and in particular, the flow separation zone can be predicted appropriately near the upstream of the flow [50]. The two equations for the computation of k and ω are as follows:

$$\frac{\partial k}{\partial t} + \nabla \cdot (\vec{u}k) - \nabla \cdot (\nu + \sigma_k \nu_T)\nabla k = P_{\omega k} - \beta^* k\omega \tag{6}$$

$$\frac{\partial \omega}{\partial t} + \nabla \cdot (\vec{u}\omega) - \nabla \cdot (\nu + \sigma_\omega \nu_T)\nabla \omega = \alpha S^2 - \beta\omega^2 - (F_1 - 1)CD_{k\omega} \tag{7}$$

$$F_1 = \tanh\left\{\min\left[\max\left(\frac{\sqrt{k}}{\beta^*\omega z}, \frac{500\nu}{z^2\omega}\right), \frac{4\sigma_{\omega 2}k}{CD_{k\omega}z^2}\right]^4\right\} \tag{8}$$

$$CD_{k\omega} = \max\left(2\sigma_{\omega 2}\frac{1}{\omega}\frac{\partial k}{\partial x_i}\frac{\partial \omega}{\partial x_i}, 10^{-10}\right) \tag{9}$$

$$\nu_T = \frac{a_1 k}{\max(a_1\omega, SF_2)} \tag{10}$$

$$F_2 = \tanh\left\{\left[\max\left(\frac{2\sqrt{k}}{\beta^*\omega z}, \frac{500\nu}{z^2\omega}\right)\right]^2\right\} \tag{11}$$

$$\alpha_i = \alpha_1 F_1 + \alpha_2(1 - F_1) \tag{12}$$

$$\beta_i = \beta_1 F_1 + \beta_2(1 - F_1) \tag{13}$$

where F_1 and F_2 are blending functions; $CD_{k\omega}$ and $P_{\omega k}$ are the limited production functions of the turbulent viscosity; ω is the rate of dissipation of the eddies; α, β, and σ coefficients [52] are $\alpha_1 = 0.5532$, $\alpha_2 = 0.4403$, $\beta_1 = 0.075$, $\beta_2 = 0.0828$, $\beta^* = 0.09$, $\sigma_k = 0.8503$, $\sigma_{\omega 1} = 0.5$, and $\sigma_{\omega 2} = 0.8561$; $a_1 = 0.31$; and S is the absolute value of the vorticity [53]. A detailed description of the model parameters is given in [50,52].

3. Model Setup

The numerical model was set up to replicate the laboratory experiments [6], which analyzed the flow characteristics along a flume in two-dimensional fixed sand dunes. Laboratory experiments were carried out with an artificial dune model of 1.6 m length, 0.08 m height, and 1.5 m width with a glued-bed median grain size of D_{50} = 1.6 mm. The flow properties in the experiments were measured in single parts of the dunes. The results of the experiments were used to validate numerous numerical researches [22,23,36,47]. One of the experimental cases [6] was selected to evaluate the numerical calculation in this study. The conditions of the selected cases are shown in Table 1.

Table 1. Dimensions of the artificial dune and experimental conditions.

L (m)	h (m)	b (m)	D_{50} (10^{-3} m)	T (°C)
1.6	0.08	1.5	1.6	18

The computational domain in this research is a two-dimensional single dune, as in Figure 2. To improve the simulation results in the vicinity of the bottom, a mesh that became gradually finer toward the bottom was generated, as shown in Figure 2. Experimental results were validated to show grid size convergence using the k-ε model with three different sizes of Δz (first node from the wall) [22]. Therefore, we added a 1 m stabilization zone, where the number of meshes was 11,000 (200 in the x direction and 55 in the y direction) in front of the inlet boundary. For the main body of the simulation, 17,600 meshes were setup with 320 and 55 meshes in the x and y directions, respectively. The boundary conditions for seven cases of different bed roughness heights were determined, as shown in Table 2.

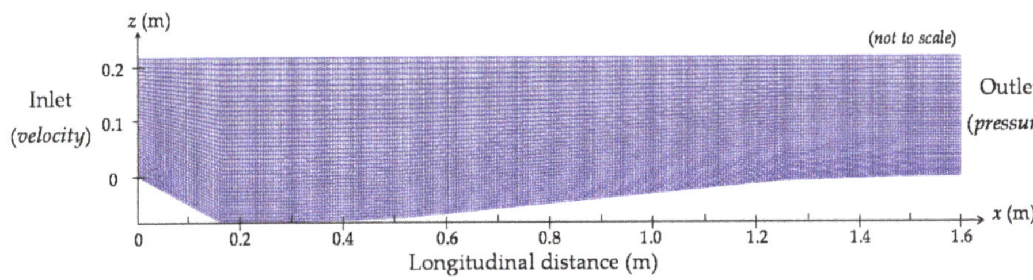

Figure 2. Computational domain of two-dimensional open-channel flow over single dune.

Table 2. Boundary conditions of turbulent models.

h_0 (m)	u_0 (m s^{-1})	k (10^{-4} m^2 s^{-2})	ε (10^{-4} m^2 s^{-3})	ω (s^{-1})	Δz (mm)
0.213	0.466	8.14	2.56	3.49	0.15

The vertical profile of the stream-wise flow velocity is controlled by the water depth, the slope, and the roughness of the wall [45]. In particular, wall roughness affects the bottom shear stress as well as the logarithmic velocity profile in uniform flow. The effect of the bed roughness is more dominant in the recirculation zone, which has been referred to as the separation zone [54,55]. The bed roughness for flow in natural rivers depends on the total induced resistance. Bed roughness can also be expressed by the roughness height. Therefore, it is necessary to set the appropriate roughness height on the wall to obtain a more accurate numerical result. We selected and applied seven experimental cases from empirical formula used mainly for open-channel cases. Values of roughness height used in the laboratory experiments [6] and values from sand grain are summarized in Table 3.

Table 3. Roughness values of cases.

Case No.	Case 1	Case 2	Case 3	Case 4	Case 5	Case 6	Case 7
Formula	D_{50}	$1.6D_{50}$	$3D_{90}$	$3.5D_{84}$	$5.2D_{65}$	$5.1D_{84}$	$6.8D_{50}$
k_s (mm)	1.6	2.5	5.55	6.3	8.84	9.18	10.88

4. Results and Discussion

4.1. Flow Velocity and Turbulence Distribution

Two turbulent modeling schemes (k-ε and k-ω SST) and seven roughness heights were considered for the simulations to find the most suitable method for determining the bed roughness properties using the turbulence model. The simulated results were compared to measured data from [6]. The numerical results with roughness height 2.5 mm (Case 2) set by the numerical modeling are shown in Figures 3–5. The profiles of the dimensionless flow velocity of the flow direction (u_x/u_0) and vertical direction (u_z/u_0) as well as the dimensionless TKE (k/u_0^2) were plotted with respect to the dimensionless vertical position (z/h_0). Overall, the profiles of the numerical results of both k-ε and k-ω SST models are in good agreement with the measured data, especially underneath $z/h_0 = 0$. However, the velocity profiles of the k-ω SST modeling results are in better agreement than those of k-ε modeling results near the bottom and the separation zone from $x = 0$ m to $x = 0.3$ m. The TKE values are defined by the mean value of kinetic energy related to the turbulence, and results of both models showed remarkable trends of turbulent intensity along the dune crest line. Numerical results of the k-ω SST model were slightly better than those of k-ε modeling overall, as analyzed by previous researches [23]. Thus, the k-ε model had a limit of not being suitable for simulating the flow in the vicinity of the channel bottom, because the model was developed for models with large Reynolds numbers to solve the flow with relatively higher turbulent regions. Therefore, wall function was applied to calculate the flow and turbulence near the bottom and this function was applied to the first cell from the bottom which was able to treat the roughness effect [23,40].

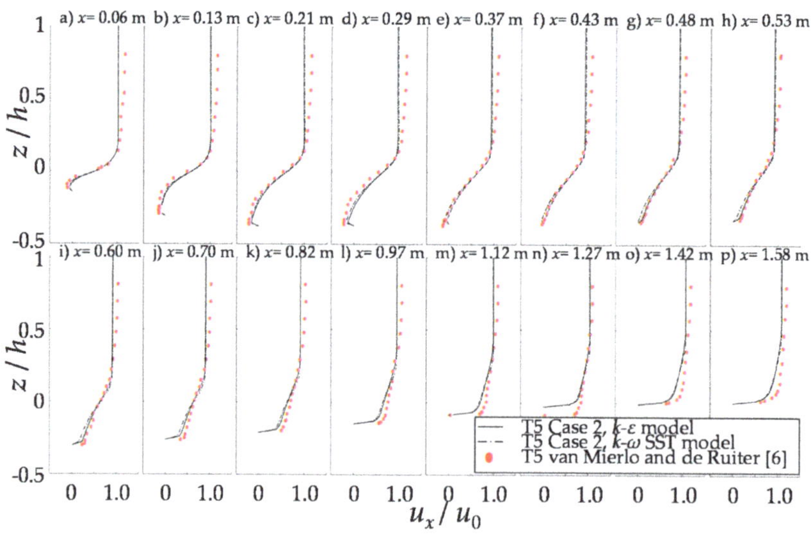

Figure 3. Comparison of different turbulence schemes in nondimensional stream-wise velocity with experimental data for Case 2 ($k_s = 1.6D_{50}$).

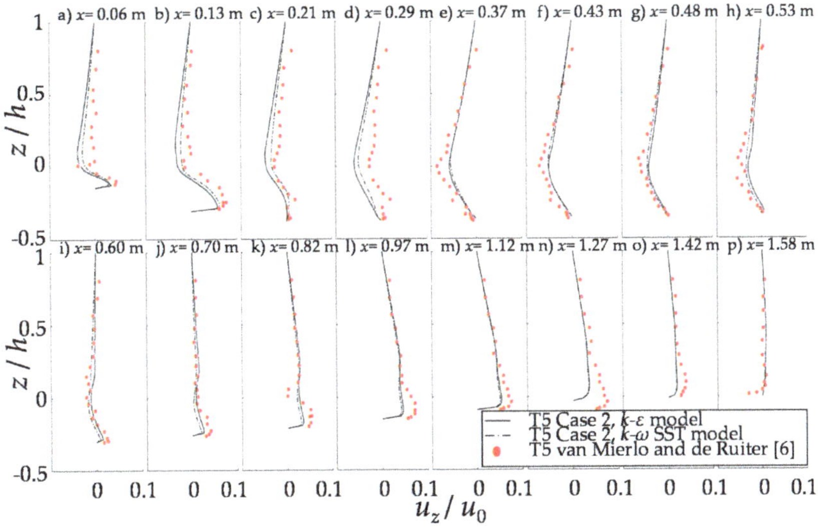

Figure 4. Comparison of different turbulence schemes in nondimensional depth-wise velocity with experimental data for Case 2 ($k_s = 1.6D_{50}$).

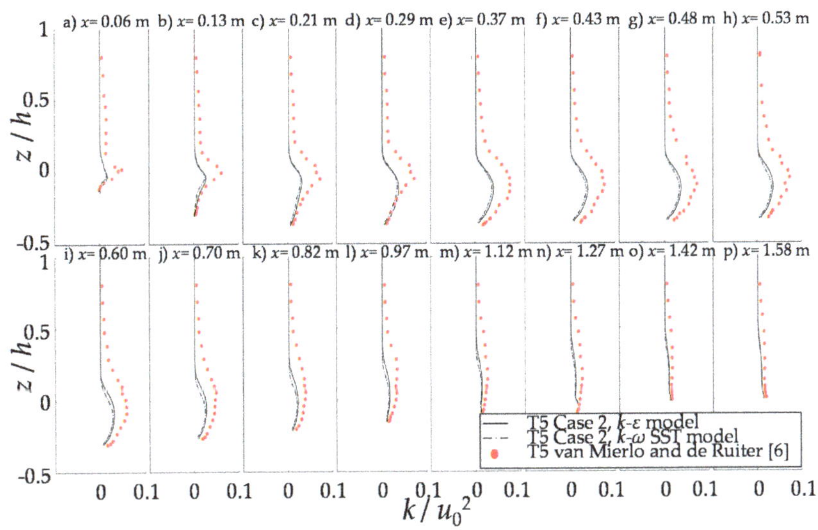

Figure 5. Comparison of different turbulence schemes in nondimensional turbulent kinetic energy (TKE) with experimental data for Case 2 ($k_s = 1.6D_{50}$).

4.2. Comparison with Experiments

Numerical results of all cases were quantified using RMSE for analysis to determine more suitable case roughness values because there was no remarkable variation in the profiles from those of Case 2, both by k-ε and k-ω SST. The results of the RMSE were expressed about two regions, i.e., from the model bottom to the free surface, and the bottom to the dune crest depth. The RMSE was calculated as follows:

$$\text{RMSE} = \sqrt{(u_{\text{mod}} - u_{\text{exp}})^2} \tag{14}$$

where u_{mod} is the numerical result and u_{exp} is measured data from the experiment. We considered all vertical measured values as well as the lower part of the water depth deeper than $z/h_0 = 0$ to focus on the roughness effect near the separation zone by comparing the RMSE values. The effect of bed roughness variation is predominant near the wall region from the bottom to approximately 80% of the dune height [54]. Measuring points with more than three measurements were selected in this analysis. Therefore, there were 16 measuring points in the experiment, 13 of which were used to represent the RMSE results (points at x = 1.27, 1.42, and 1.58 m were excluded). The distributions of two RMSE values are compared in Figures 6–8.

Overall, the averaged RMSE of the u_x for the whole water depth (Figure 6) was stable, whereas the partially averaged results (Figure 6c,d) exhibited large differences upstream and downstream in the lower part (from initial bed elevation to the deepest point) of z/h_0 where the flow velocity fluctuated remarkably. In addition, the RMSE values of Case 1 and Case 2, which had relatively small roughness height values, showed a similar tendency. Case 3 to Case 7, which had greater roughness height values, exhibited identical patterns (Figure 6). The RMSE results of the upstream part are also less than those at the downstream. All water-depth averaged RMSE values were generally similar to those at the downstream direction (Figure 6a,b). However, the RMSE values in the lower part, from Case 3 to Case 7, increased at the downstream part of the dune (Figure 6c,d). This pattern is consistent with the fact that the dune geometry is a dominant factor of the entire depth of flow and that the effect of the bed roughness is predominant in the separation [54,55]. The RMSE values of both turbulence models in the lower part are smaller than the RMSE values of the entire water depth, especially at the upstream, and the RMSE of the k-ω SST model is slightly smaller than that of the k-ε model. In particular, the smallest value of RMSE with the k-ω SST model is smaller than that of the k-ε model, and the point changed from x = 0.43 m in the k-ε model to x = 0.29 m in the k-ω SST. model (Figure 6c,d). The fully water-depth averaged values of RMSE with both turbulence models were estimated to be 0.0385–0.0630 and 0.0392–0.0613 m/s, respectively. Moreover, the values in the lower part were estimated to be 0.003–0.1416 m/s for the k-ε model and 0.0072–0.0152 m/s for the k-ω SST model. The smallest values of RMSE in the lower vicinity of the dune upstream were 0.0111, 0.0124, and 0.0135 m/s in Case 3 of k-ω SST, at x = 0.21 m, respectively. The suggested combination of turbulence model and initial condition for the roughness height in the inlet part were evaluated for each focus of analysis such as overall flow and turbulence distributions, as well as regional backwater trend at the downstream of the dune inlet.

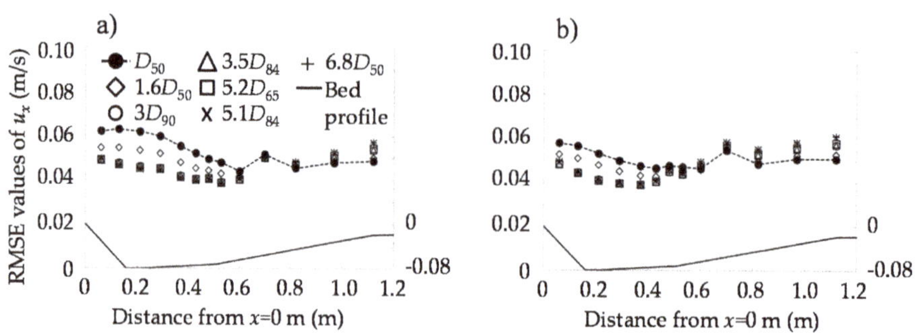

Figure 6. *Cont.*

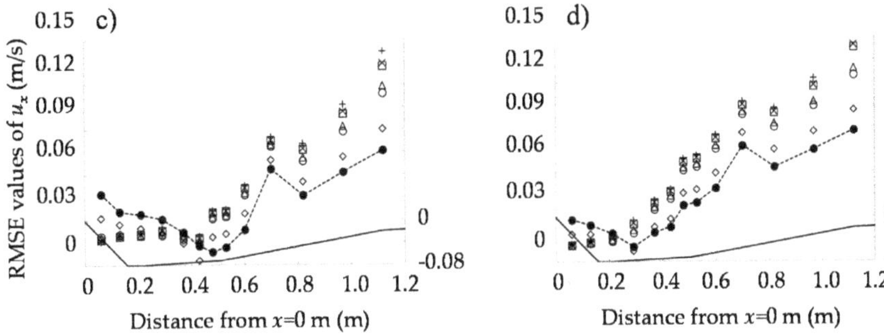

Figure 6. Root mean square error (RMSE) values of flow velocity in stream-wise direction with various bottom roughness height formulas: (**a**) Fully depth-averaged ($0 < z/h_0 < 1$) RMSE with k-ε model; (**b**) Fully depth-averaged with k-ω shear-stress transport (SST) model; (**c**) Partially depth-averaged ($0 < z/h_0 < 0.15$) k-ε model for the lower part; (**d**) Partially depth-averaged ($0 < z/h_0 < 0.15$) k-ω SST model.

The longitudinal distribution of the RMSE values of u_z are plotted in Figure 7. The overall RMSE trends are similar for both turbulence models. Higher roughness value cases (Cases 3–7) showed better fit than lower cases (Cases 1 and 2). For the k-ε model (Figure 7a,c), the RMSE values at the upstream part up to x = 0.13 m were larger than at the downstream part. The depth-averaged RMSE values of the two turbulence models were estimated to be 0.0028–0.0148 and 0.0029–0.0120 m/s, respectively, and the RMSE values in the lower part near the upstream up to x = 0.21 m were estimated to be 0.0078–0.0156 and 0.0058–0.0114 m/s for the k-ε and k-ω SST models, respectively.

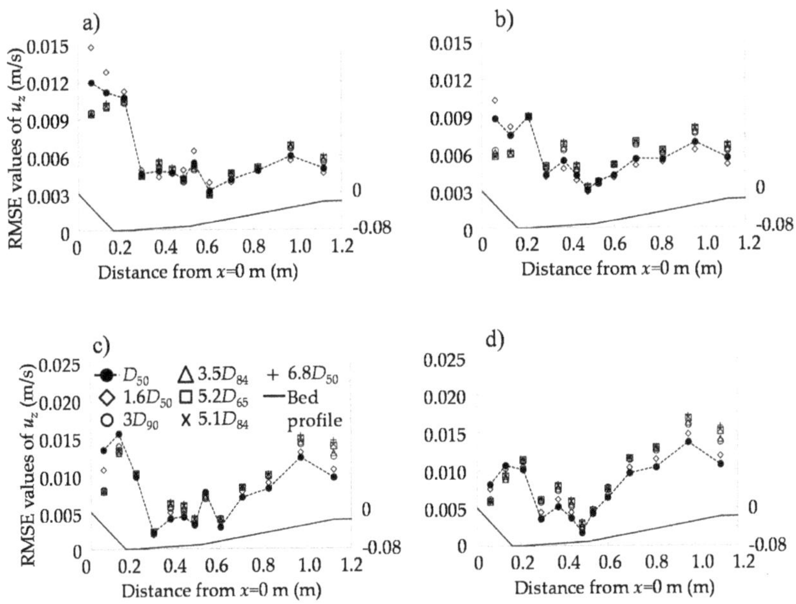

Figure 7. RMSE of flow velocity in depth-wise direction with various bottom roughness height formulas: (**a**) Fully depth-averaged ($0 < z/h_0 < 1$) RMSE with k-ε model; (**b**) Fully depth-averaged with k-ω SST model; (**c**) Partially depth-averaged ($0 < z/h_0 < 0.15$) k-ε model for the lower part; (**d**) Partially depth-averaged ($0 < z/h_0 < 0.15$) k-ω SST model.

The RMSE values of the TKE results about all cases are shown in Figure 8. The RMSE values were estimated to be relatively large downstream of the separation zone. These results were reflected due to the high turbulence in the wake region arising directly behind the flow separation zone [15] based on a comparison with the experimental data in Figure 3. Larger roughness cases (Cases 3–7) have substantially lower depth-averaged RMSE values than the smaller roughness cases for the entire application area, as shown in Figure 8a,b. The RMSE values in the lower part (Figure 8c,d) were similar for all cases to the wake region, and the RMSE values of the larger roughness cases had a better fit at the downstream. Theoretically, the *k*-*ε* model should have the disadvantage of inaccurate flow properties for simulations near the wall region because this model was designed for large Reynolds number cases to solve fully developed turbulent flow. However, the results from the two models showed minor differences because the wall function was applied to overcome the limitation and to calculate the roughness effect, as mentioned previously [23]. In this study, the RMSE values of the TKE in the lower part were difficult to use for comparison of various roughness cases, as shown in Figure 8c,d. The characteristics of TKE are not representative of the flow near the wall because they are large below the dune crest ($x = 0$), and also above the crest (Figure 4). A previous research by [54] also revealed that the peak values of the turbulent properties were located in the dune crest line from the channel bottom which was rarely affected by the bed roughness. Consequently, we selected the optimal roughness case from the results of the flow velocities in the *x*-*z* plane, and $3D_{90}$ (=Case 3) proposed by [56] was chosen for the focus on the separation zone of the dune.

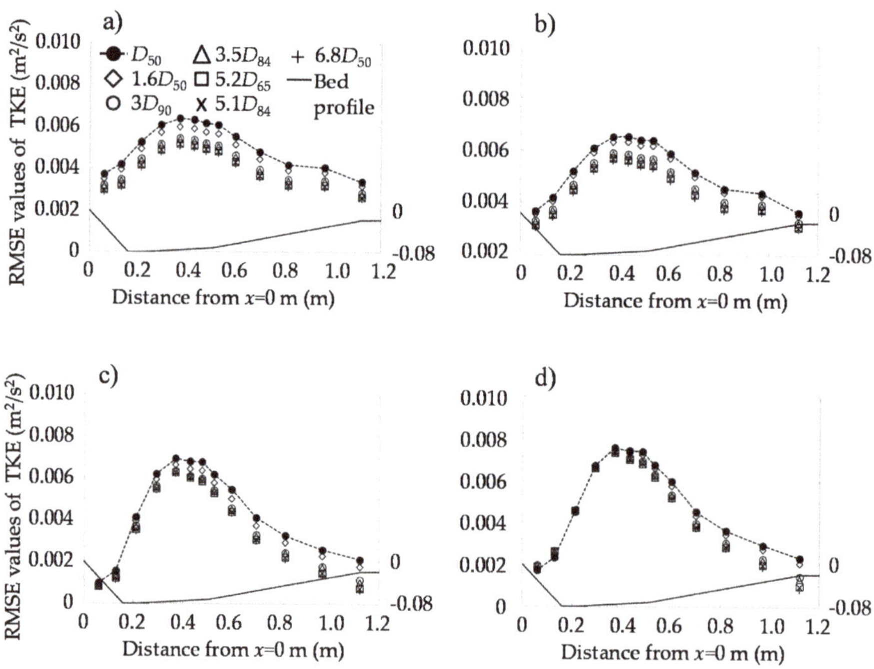

Figure 8. RMSE of TKE with various bottom roughness height formulas: (**a**) Fully depth-averaged ($0 < z/h_0 < 1$) RMSE with *k*-*ε* model; (**b**) Fully depth-averaged with *k*-*ω* SST model; (**c**) Partially depth-averaged ($0 < z/h_0 < 0.15$) *k*-*ε* model for the lower part; (**d**) Partially depth-averaged ($0 < z/h_0 < 0.15$) *k*-*ω* SST model.

For overall analysis, in Figure 9, flow velocity and turbulent kinetic energy data in Case 3 with *k*-*ε* and *k*-*ω* SST models were compared with respect to the stream-wise location (Figure 9). In the figure, numerical results of the stream-wise flow velocity component were rather underestimated over the entire section for all depth directions except for the lower part near the inlet where backwater occurred. Nevertheless, the depth-wise velocity showed a distribution that was somewhat consistent with the experimental results, except for several outliers in Figure 9c,d. In addition, a comparison of the TKE data showed that the numerical results were underestimated in Figure 9e,f. These results are considered to be a limitation of the RANS model application, and in the future, it can be improved by applying models such as LES. Nevertheless, the combination of the RANS model and roughness height found in this study secured applicability, because it was possible to adequately simulate the backwater generated in the lower part.

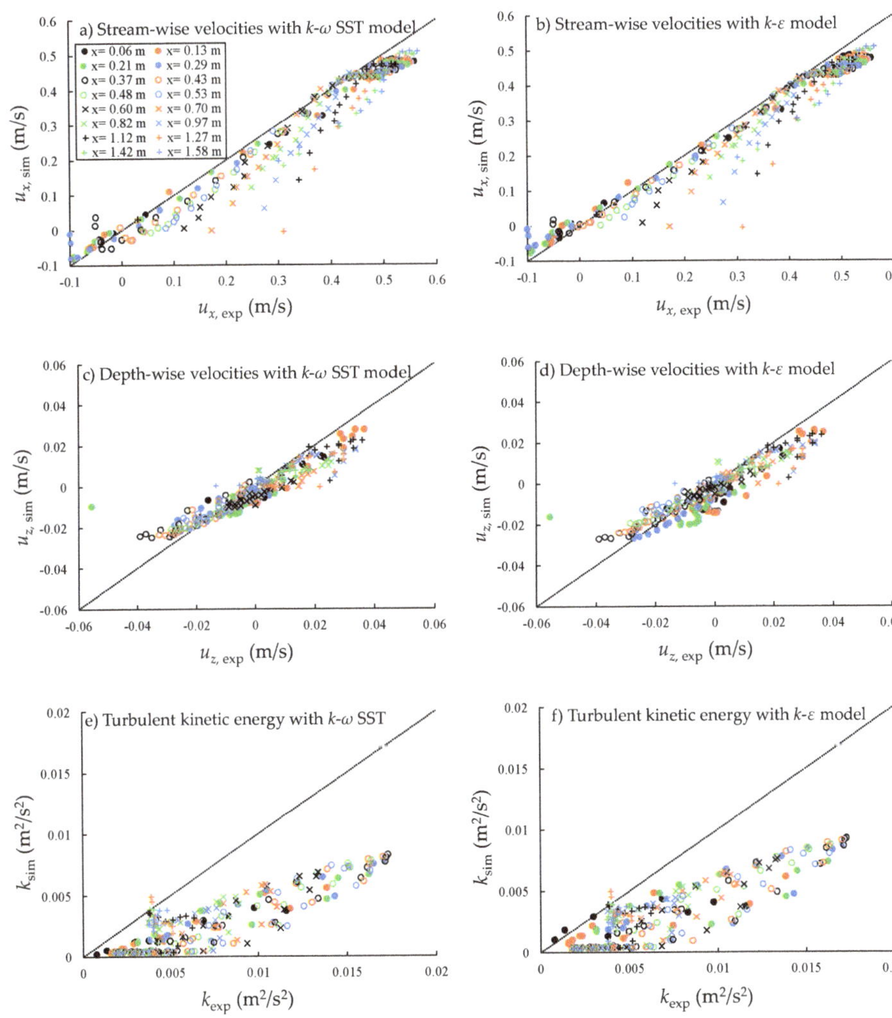

Figure 9. Comparison of simulated and experimental data of flow velocity and TKE with perfect line of agreement (= dashed line) of Case 3 ($k_s = 3D_{90}$).

5. Summary and Conclusions

This study focused on finding the appropriate numerical approach with roughness boundary formula at the channel bottom and turbulence scheme for modeling the flow and turbulence characteristics in an artificial dune. Roughness values from seven formula for use in two turbulent models, k-ε and k-ω SST, were compared. The simulated results of the flow velocity and TKE profiles were compared with the physical experimental data from [6], and the calculated results of 14 combined cases were also compared using the estimated RMSE values. The flow velocity profiles and TKE results from the k-ε and k-ω SST models were calculated very similarly throughout the flow area in the artificial dune. Additionally, variations in roughness height affected the results of the turbulent modeling. Cases with higher roughness heights proposed previously were more accurate than those with lower values. However, in this study, we tried to find the appropriate combination of numerical modeling in the vicinity of the channel bottom. Therefore, we estimated the RMSE values in the separated zone for the lower and upper part from the initial water surface elevation and the overall part. In particular, the RMSE values of stream-wise velocity in the vertically lower part were remarkably lower than the fully depth-wise averaged RMSE values in the vicinity of the upstream part of the single dune. However, the overall RMSE values from the inlet to $x = 0.29$ m were the most accurate in Case 3. In terms of the depth-wise directional velocity, the results of the k-ω SST model upstream part were more accurate and the RMSE values in the wake region were smaller. Consequently, we determined that the combination of the k-ω SST model and Case 3 ($k_s = 3D_{90}$) was the most suitable for the research focus of the flow separation zone. Therefore, considering the accuracy of the whole study area, both turbulence models could be applied for modeling downstream of the separation zone with a wall function. On the basis of the combination suggested in this study, it is expected that more accurate combinations could be found in the future by supplementing additional turbulence analysis modules and adding behavior simulations in the width-direction geometry of the model.

Author Contributions: Conceptualization, J.A. and S.W.P.; methodology, J.L.; software, J.L.; validation, J.L., S.W.P., and J.A.; formal analysis, J.L.; investigation, J.L. and S.W.P.; resources, S.W.P.; data curation, J.L.; writing—original draft preparation, J.L.; writing—review and editing, J.A.; visualization, J.L. and S.W.P.; supervision, J.A.; project administration, J.A.; funding acquisition, J.A. All authors have read and agreed to the published version of the manuscript.

Funding: This research received no external funding.

Acknowledgments: This work was supported by Incheon National University Research Grant in 2018.

Conflicts of Interest: The authors declare no conflict of interest.

Notations

α_i,β_i,σ_i	Model closure coefficients ($\alpha_1 = 0.5532$, $\alpha_2 = 0.4403$, $\beta_1 = 0.075$, $\beta_2 = 0.0828$, $\sigma_k = 0.8503$, $\sigma_{\omega 1} = 0.5$, and $\sigma_{\omega 2} = 0.8561$)
β^*	Empirical constant (=0.09)
F_1, F_2	Blending functions
a_1	Calibration coefficient (=0.31)
$CD_{k\omega}$	Limited production function of the turbulent viscosity
D_n	Diameter of particle intermediate axis for which n = 50%, 65%, 84%, and 90%, respectively, of the sample of bed material is finer
F_2	Second blending function
h	Dune height
h_0	Inlet water depth
k	Turbulent kinetic energy
k_s	Roughness height
L	Dune length
b	Dune width
p	Pressure
P_k	Turbulent viscosity production
$P_{\omega k}$	Limited production term of the turbulent viscosity
S	Absolute value of vorticity
u_x	Flow velocities in stream-wise direction

u_z	Flow velocities in water-depth direction
u_0	Flow velocity at inlet
u_{exp}	Observed values (flow velocity and turbulent kinetic energy) from experiments in the previous research
u_{mod}	Calculated values (flow velocity and turbulent kinetic energy) from numerical modeling
$\vec{u}$	Velocity vector
x	Stream-wise direction
z	Depth-wise directional distance
t	time
Δz	Mesh size of first node from the wall
ε	Turbulent dissipation
ν	Kinematic viscosity coefficient
ν_T	Turbulent kinematic eddy viscosity coefficient
ω	Turbulent frequency (rate of dissipation of the eddies)
T	Water temperature

References

1. Breusers, H.N.C. Time scale of two-dimensional local scour. In Proceedings of the 12th IAHR-Congress, Fort Collins, CO, USA, 11–14 September 1967; pp. 275–282.
2. Hoffmans, G.J.C.M.; Booij, R. Two-dimensional modelling of local-scour holes. *J. Hydraul. Res.* **1993**, *31*, 615–634. [CrossRef]
3. Park, S.W. Experimental Study of Local Scouring at the Downstream of River Bed Protection. Ph.D. Thesis, Seoul National University, Seoul, Korea, February 2016.
4. Park, S.W.; Hwang, J.W.; Ahn, J. Physical modeling of spatial and temporal development of local scour at the downstream of bed protection for low Froude number. *Water* **2019**, *11*, 1041. [CrossRef]
5. Ahn, J.; Yen, H. Semi-two dimensional numerical prediction of non-equilibrium sediment transport in reservoir using stream tubes and theory of minimum stream power. *KSCE J. Civ. Eng.* **2015**, *19*, 1922–1929. [CrossRef]
6. van Mierlo, M.C.L.M.; de Ruiter, J.C.C. *Turbulence Measurements above Artificial Dunes*; Report No. Q789; Delft Hydraulic Laboratory: Delft, The Netherlands, 1988.
7. Lyn, D.A. Turbulence measurements in open-channel flows over artificial bed forms. *J. Hydraul. Eng.* **1993**, *119*, 306–326. [CrossRef]
8. Nelson, J.M.; Mclean, S.R.; Wolfe, S.R. Mean flow and turbulence fields over two-dimensional bed forms. *Water Resour. Res.* **1993**, *29*, 3935–3953. [CrossRef]
9. Mclean, S.R.; Nelson, J.M.; Wolfe, S.R. Turbulence structure over two-dimensional bed forms: Implications for sediment transport. *J. Geophys. Res.* **1994**, *99*, 12729–12747. [CrossRef]
10. Bennett, S.J.; Best, J.L. Mean flow and turbulence structure over fixed, two-dimensional dunes: Implications for sediment transport and bed form stability. *Sedimentology* **1995**, *42*, 491–513. [CrossRef]
11. Coleman, S.E.; Melville, B.W. Initiation of bed forms on a flat sand bed. *J. Hydraul. Eng.* **1996**, *122*, 301–310. [CrossRef]
12. Kadota, A.; Nezu, I. Three-dimensional structure of space-time correlation on coherent vortices generated behind dune crest. *J. Hydraul. Res.* **1999**, *37*, 59–80. [CrossRef]
13. Maddux, T.B. Turbulent Open Channel Flow over Fixed Three-Dimensional Dune Shapes. Ph.D. Thesis, University of California, Santa Barbara, CA, USA, 2002.
14. Kostaschuk, R.; Villard, P.; Best, J. Measuring velocity and shear stress over dunes with acoustic Doppler profiler. *J. Hydraul. Eng.* **2004**, *130*, 932–936. [CrossRef]
15. Best, J. The fluid dynamics of river dunes: A review and some future research directions. *J. Geophys. Res.* **2005**, *110*, F04S02. [CrossRef]
16. Best, J. Kinematics, topology and significance of dune-related macroturbulence: Some observations from the laboratory and field. *Spec. Publs. Int. Ass. Sediment* **2005**, *35*, 41–60. [CrossRef]
17. Coleman, S.E.; Nikora, V.I. Fluvial dunes: Initiation, characterization, flow structure. *Earth Surf. Process. Landf.* **2011**, *36*, 39–57. [CrossRef]

18. Lee, H.Y.; Hsieh, H.N. Numerical simulations of scour and deposition in a channel network. *Int. J. Sediment Res.* **2003**, *18*, 32–49.
19. Nagata, N.; Hosoda, T.; Nakato, T.; Muramoto, Y. Three-dimensional numerical model for flow and bed deformation around river hydraulic structures. *J. Hydraul. Eng.* **2005**, *131*, 1074–1087. [CrossRef]
20. Wu, C.L.; Chau, K.W. Mathematical model of water quality rehabilitation with rainwater utilization: A case study at Haigang. *Int. J. Environ. Pollut.* **2006**, *28*, 534–545. [CrossRef]
21. Liu, X.; Garcia, M.H. Three-dimensional numerical model with free water surface and mesh deformation for local sediment scour. *J. Waterw. Portcoastaland Ocean Eng.* **2008**, *134*, 203–217. [CrossRef]
22. Guven, A.; Gunal, M. Hybrid modelling for simulation of scour and flow patterns in laboratory flumes. *Int. J. Numer. Methods Fluids* **2010**, *62*, 291–312. [CrossRef]
23. Nguyen, T.H.H.; Lee, J.; Park, S.W.; Ahn, J. Two-dimensional numerical analysis on the flow and turbulence structures in artificial dunes. *KSCE J. Civ. Eng.* **2018**, *22*, 4922–4929. [CrossRef]
24. Kyrousi, F.; Leonardi, A.; Roman, F.; Armenio, V.; Zanello, F.; Zordan, J.; Juez, C.; Falcomer, L. Large eddy simulations of sediment entrainment induced by a lock-exchange gravity current. *Adv. Water Resour.* **2018**, *114*, 102–118. [CrossRef]
25. Ouro, P.; Juez, C.; Franca, M. Drivers for mass and momentum exchange between the main channel and river bank lateral cavities. *Adv. Water Resour.* **2020**, 137. [CrossRef]
26. Pope, S.B. *Turbulent Flows*; Cambridge University Press: Cambridge, UK, 2000.
27. Hirsch, C. *Numerical Computation of Internal and External Flows: The Fundamentals of Computational Fluid Dynamics*, 2nd ed.; Butterworth-Heinemann: Massachusetts, MA, USA, 2007.
28. Bayon, A.; Valero, D.; García-Bartual, R.; Vallés-Morán, F.J.; López-Jiménez, P.A. Performance assessment of OpenFOAM and FLOW-3D in the numerical modeling of a low Reynolds number hydraulic jump. *Environ. Model. Softw.* **2016**, *80*, 322–335. [CrossRef]
29. Spalart, P.R. Strategies for turbulence modelling and simulations. *Int. J. Heat Fluid Flow* **2000**, *21*, 252–263. [CrossRef]
30. Mendoza, C.; Shen, H.W. Investigation of turbulent flow over dunes. *J. Hydraul. Eng.* **1990**, *116*, 459–477. [CrossRef]
31. Johns, B.; Soulsby, R.L.; Xing, J. A comparison of numerical model experiments of free surface flow over topography with flume and field observations. *J. Hydraul. Res.* **1993**, *31*, 215–228. [CrossRef]
32. Yoon, J.Y.; Patel, V.C. Numerical model of turbulent flow over sand dune. *J. Hydraul. Eng.* **1996**, *122*, 10–18. [CrossRef]
33. Cheong, H.F.; Xue, H. Turbulence model for water flow over two-dimensional bedforms. *J. Hydraul. Eng.* **1997**, *123*, 402–409. [CrossRef]
34. Tan, C.A.; Sinha, S.K.; Ettema, R. Ice-cover influence on near-field mixing in dune-bed channel: Numerical simulation. *J. Cold Reg. Eng.* **1999**, *13*, 1–20. [CrossRef]
35. Dimas, A.A.; Fourniotis, N.T.; Vouros, A.P.; Demetracopoulos, A.C. Effect of bed dunes on spatial development of open channel flow. *J. Hydraul. Res.* **2008**, *46*, 802–813. [CrossRef]
36. Fourniotis, N.T.; Toleris, N.E.; Dimas, A.A.; Demetracopoulos, A.C. Numerical computation of turbulence development in flow over sand dunes. In *Advances in Water Resources and Hydraulic Engineering*; Springer: Berlin/Heidelberg, Germany, 2009.
37. Hoffmans, G.J.C.M.; Verheij, H.J. *Scour Manual*, 1st ed.; CRC Press/A.A Balkema: Rotterdam, The Nethelands, 1997. [CrossRef]
38. Amoudry, L.O.; Liu, P.L.F. Two-dimensional, two phase granular sediment transport model with applications to scouring downstream of an apron. *Coast. Eng.* **2009**, *56*, 693–702. [CrossRef]
39. Termini, D. Bed scouring downstream of hydraulic structures under steady flow conditions: Experimental analysis of space and time scales and implications for mathematical modeling. *Catena* **2011**, *84*, 125–135. [CrossRef]
40. Nguyen, T.H.H.; Ahn, J.; Park, S.W. Numerical and physical investigation of the performance of turbulence modeling schemes around a scour hole downstream of a fixed bed protection. *Water* **2018**, *10*, 103. [CrossRef]
41. Spalart, P.R.; Allmaras, S.R. A one-equation turbulence model for aerodynamic flows. In Proceedings of the 30th Aerospace Sciences Meeting and Exhibit (AIAA, 92–0439), Reno, NV, USA, 6 January 1992; p. 439. [CrossRef]

42. Colebrook, C.F.; White, C.M. Experiments with fluid friction in roughened pipes. *Proc. R. Soc. A Math. Phys. Eng. Sci.* **1937**, *161*, 367–381. [CrossRef]
43. Nelson, J.M.; Smith, J.D. Mechanics of flow over ripples and dunes. *J. Geophys. Res.* **1989**, *94*, 8146–8162. [CrossRef]
44. Jellesma, M. Form Drag of Subaqueous Dune Configurations. Master's Thesis, University of Twente, Enschede, The Netherlands, November 2013.
45. van Rijn, L.C. *Principles of Sediment Transport in Rivers, Estuaries, and Coastal Seas*; Aqua Publications: Blokzijl, The Netherlands, 1993.
46. McLean, S.R.; Wolfe, S.R.; Nelson, J.M. Spatially averaged flow over a wavy boundary revisited. *J. Geophys. Res.* **1999**, *104*, 15743–15753. [CrossRef]
47. Nelson, J.M.; Burman, A.R.; Shimizu, Y.; McLean, S.R.; Shreve, R.L.; Schmeeckle, M. Computing flow and sediment transport over bedforms. *Rivercoastal Estuar. Morphodynamics* **2005**, *2*, 861–872. [CrossRef]
48. Zhou, T.; Endreny, T.A. Reshaping of the hyporheic zone beneath river restoration structures: Flume and hydrodynamic experiments. *Water Resour. Res.* **2013**, *49*, 5009–5020. [CrossRef]
49. Launder, B.E.; Spalding, D.B. The numerical computation of turbulent flows. *Comput. Methods Appl. Mech. Eng.* **1974**, *3*, 269–289. [CrossRef]
50. Menter, F.R. Two-equation eddy-viscosity turbulence models for engineering applications. *AIAA J.* **1994**, *32*, 1598–1605. [CrossRef]
51. Wilcox, D.C. Re-assessment of the scale-determining equation for advanced turbulence models. *AIAA J.* **1988**, *26*, 1299–1310. [CrossRef]
52. Menter, F.R.; Kuntz, M.; Langtry, R. Ten years of industrial experience with the SST turbulence model. In Proceedings of the 4th International Symposium On Turbulence, Heat And Mass Transfer, Antalya, Turkey, 12–17 October 2003; Volume 4, pp. 625–632.
53. Robertson, E.; Choudhury, V.; Bhushan, S.; Walters, D.K. Validation of OpenFOAM numerical methods and turbulence models for incompressible bluff body flows. *Comput. Fluids* **2015**, *123*, 122–145. [CrossRef]
54. Balachandar, R.; Patel, V.C. Flow over a fixed dune. *Can. J. Civ. Eng.* **2008**, *35*, 511–520. [CrossRef]
55. Balachandar, R.; Reddy, H.P. *Bed Forms and Flow Mechanisms Associated with Dunes, Sediment Transport-Flow and Morphological Processes*; InTech: London, UK, 2011.
56. van Rijn, L.C. Sediment transport, Part III: Bed forms and alluvial roughness. *J. Hydraul. Eng.* **1984**, *110*, 1733–1754. [CrossRef]

MDPI
St. Alban-Anlage 66
4052 Basel
Switzerland
Tel. +41 61 683 77 34
Fax +41 61 302 89 18
www.mdpi.com

Water Editorial Office
E-mail: water@mdpi.com
www.mdpi.com/journal/water

www.ingramcontent.com/pod-product-compliance
Lightning Source LLC
Chambersburg PA
CBHW061602210926
51850CB00009B/2187
9783036555973